Advances in Forestry Research in India

Advances in Forestry Research in India

Praveen Taank

CYBER TECH PUBLICATIONS
4264/3, Ansari Road, Daryaganj, New Delhi-110002 (India)
Ph.: 011-23244078. Fax: 011-23280028
E-Mail: cyberpublicationsdelhi@yahoo.com

ADVANCES IN FORESTRY RESEARCH IN INDIA

Praveen Taank

First Edition 2009

Published by :

G.S. Rawat for Cyber Tech Publications
4264/3, Ansari Road, Daryaganj, New Delhi-110002 (India)
Ph.: 011-23244078. Fax: 011-23280028
E-Mail: cyberpublicationsdelhi@yahoo.com

Printed at : **Amit Enterprises**, Maujpur, Delhi, (9810878184)

Preface

This National Forest Strategy is meant to guide India's efforts in sustainable forest management as we enter a new millennium. It is a renewed plan of action to deal in a forthright manner with the connectedness among the ecological, economic, social and cultural aspects of forest use and conservation. To understand the current stage of evolution in setting forest policy at the national level, we need to review recent milestones that mark Indian forestry. This century has seen unequalled acceleration in human population, widespread industrialization, globalization of markets, the advent of the information age and the emergence of a global economy based on both material and informational resources.

In contrast to the globalization of the economy, fundamental ideological rifts are occurring worldwide between cultures which suggest or perhaps simply remind us of the overarching importance of reconciling economic pursuits with human values and relations, and with nature's ability to sustain human activity.

In the present work the team has researched deeply and largely throughout the country for the forest, flora and fauna etc. Undersigned is confident enough that his work will be appreciated by the readers.

—Author

Contents

1

Training and Research

INTRODUCTION

A number of newly independent states were added to the world community in the last decade of the twentieth century, following the split of the Soviet Union. Those countries with economies in transition occupy a large portion of the Eurasian continent, extending from Central and Eastern Europe, Central Asia, and Russia to the Baltic.

Although the geographic and environmental conditions of the countries vary significantly, they all share similar problems of political, social, economic and cultural natures in the transition process from a centrally controlled economy to a market economy. Such a transformation process has inevitable impacts on their legal and social conditions as well as on their natural environment.

The negative legacy of the planned economy is evident in the field of natural resources management in those countries, and not least in terms of sustainable forest management. In most of the transition countries, the forestry sector faces difficulties with limited financial resources and other support from the current government.

Consequently, the forest resources of many of those countries are not under proper and sustainable management practices. Meanwhile, pressures on the remaining forest resources are mounting in the course of economic development, as in the case with many other parts of the world.

NATIONAL FOREST PROGRAMMES

At the international level, there has been an initiative to build a solid institutional framework and guidelines for the forest-sector-based development in all countries that deal with forests directly or indirectly in various forms of protection and utilization. For this purpose, under the auspices of the United Nations Forum on Forests (UNFF), the Intergovernmental Panel on Forests and the Intergovernmental Forum on Forests (IPF/IFF) drew up a set of proposals for action. In this, establishment, implementation, monitoring and evaluation of national forest programmes (NFPs) are strongly encouraged for all countries as a framework for promoting sustainable forest management. A number of organizations acting in forest sector development in different parts of the world, including major international organizations such as FAO, ITTO, and the World Bank as well as major bilateral donor agencies, have spent some ten years in designing NFPs.

Today, NFPs are recognized as the major coordinating policy development framework to promote sustainable management and conservation of forests, under the auspices of the UN Commission on Sustainable Development. One of the main principles of the NFP processes is to contribute not only to the forestry sector, but also to the overall sustainable development strategies. Therefore, it requires integration into each country's overall socio-economic development plans. In other words, by facilitating development of the intellectual, human and institutional capacity, NFPs aim to embody an enabling environment for sustainable forest management as well as for forest-based economic and social development. Thus, the IPF/IFF proposals urge countries-particularly developing states and countries with economies in transition-to include capacity building as an objective of their NFPs, with particular attention to training, extension services, technology transfer and financial assistance from developed countries.

It is now seen as an imminent task for the countries with economies in transition to establish appropriate NFPs under the initiative of their respective governments, in order to achieve sustainable management of their forest resources. Especially urgent and essential for establishing a sound basis for sustainable forest management is integration of research and education into the development of appropriate forest policies (NFPs) and practices, alongside the necessary institution building. In a transitional society facing numerous challenges related to prioritization in overall resources allocation and mobilization, there is a lack of crucial prerequisites for forest policy development processes, including the research and education capacity of the forestry sector. This means not only a shortage of financial and human resources, but also fundamental problems pertaining to the education and academic research systems and facilities.

Therefore, capacity building in the interface of forest research and education is a serious challenge for forest policy development in countries with economies in transition. However, only limited attention has thus far been paid to the need for establishing linkages between forest policy, research, and education. This is related to the lack of international coordination in developing forest research and academic education at the global level-a process which this project aims to facilitate.

The current project, initiated by the United Nations University (UNU), European Forest Institute (EFI) and the University of Joensuu-SILVA Network, tries to open the way out of such a deadlocked situation. For this purpose, the project will first examine the state-of-the-art of and challenges for forest research and higher education in the countries with economies in transition, and analyze the possibilities of research and education to contribute to forest policy development processes in those countries. Based on the findings of the thorough evaluation and analysis of the current situation in the region and in each country, recommendations will be made regarding specific actions to be taken.

It is expected that the required actions will vary from one country to another, depending on the level of economic and social development, political and sectoral commitment, needs and conditions of forest resources, etc. In many cases, however, the problems and challenges would also have similar features, and it is expected that collective learning from each other's experience can lead to more efficient alleviation of the current situation. To enable those countries to share their experience and expertise, the project also aims to support development of forest-related regional networks and research projects in and among the 28 targeted countries with economies in transition.

To fulfill the above objectives, a series of joint conferences and seminars are planned that will bring together various stakeholders of the forestry sector in countries with economies in transition, including policy makers, public and private entities operating in the forestry sector, and forest-related university faculties, researchers and non-governmental organizations. International organizations, research institutions, donor agencies, and governmental and non-governmental organizations from outside the transition countries are also invited to strengthen the networks and create partnership arrangements. The first of such conferences took place in July 2002 in Kaunas, Lithuania. Training workshops are also to be held for the purpose of capacity development of those actively engaged in forest management in the targeted countries. The project organizes "summer schools" to promote discussion on forest sciences in the context of national and international forest policy dialogue, and plans to develop M.Sc. courses on policy analysis and forest economics. Also in the project scope is further education for forestry administration and forest policy planning.

Information on the findings and outcomes of the above conferences, seminars and workshops will be disseminated to all those concerned and to the general public through publications and websites.

PROJECT TARGET COUNTRIES

United Nations University (UNU)

UNU is an organ of the United Nations established by the General Assembly in 1973. Its objective is to form an international community of scholars engaged in research, advanced training and the dissemination of knowledge related to the pressing global problems of human survival, development and welfare. Its activities focus mainly on peace and conflict resolution, environment and sustainable development in the changing world, and science and technology in relation to these issues. As a think-tank of the UN system, the University operates a worldwide network of research and post-graduate training centres, with its planning and coordinating headquarters in Tokyo, Japan. Since the mid-1990s, the Environment and Sustainable Development (ESD) Unit of UNU has been working actively on the issue of sustainable forest development within the framework of the World Forest, Society and Environment (WFSE) Research Programme. The mission of this programme is to conduct globally relevant research and human capacity building in forests, society and environment in support of sustainable forest management and the well-being of people.

European Forest Institute (EFI)

EFI is a leading forest research network in Europe with its hub in Joensuu, Finland. It has currently 142 members from 39 countries, which are mostly forest research organizations, universities, and industry and non-governmental organizations. EFI's mission is to promote, conduct and cooperate in research of forestry and forest products at the pan-European level, and to make the results of the research known to all interested parties, notably in the areas of policy formulation and implementation. It thus aims to promote the conservation and sustainable management of forests for producing goods and services in a sustainable way. With its unique network of forest scientists and other professionals, EFI conducts and facilitates research in such priority areas as forest ecology and management, forest product markets and socio-economics, policy analysis, and forest resources and information.

GUIDE FOR SURFACE COAL MINE RECLAMATION PLANS

Coal has been mined in Nova Scotia for over 300 years. Surface and underground coal operations have made significant economic contributions to the province, including fuel, steel manufacturing and electrical generation. Excavation is the primary means for recovering coal resources, which produces environmental disturbances. The mining industry in Nova Scotia must continue to adapt to new standards and regulatory requirements for responsible development of mineral resources, which includes integrating environmental issues with economic planning.

Mine reclamation is an integral part of the mineral development process. It is designed to restore to an acceptable state the physical, chemical, and biological quality of land and water regimes disturbed by mining. This 'Guide for Surface Coal Mine Reclamation Plans' has been developed by Nova Scotia Environment and Labour in collaboration with the Department of Natural Resources. It includes best management practices for surface mine reclamation within Nova Scotia's coal mining industry. It will also provide communities with greater certainty regarding sustainable environmental practices with respect to reclamation of surface coal mines. A glossary of terms used in this guide is provided at the end of the document.

APPLICATION

Both the Department of Natural Resources, and Environment and Labour, are accountable for specific responsibilities under the Mineral Resources Act and the Environment Act regarding reclamation of surface coal mines. Both Departments share responsibility for review and approval of surface coal mine reclamation plans submitted by industry. Common guidance is appropriate and should assist regulatory reviews by both agencies. Specifically, this guidance is intended to provide a framework for industry to use in their reclamation planning and during environmental assessments, industrial approval processing and submissions to the Departments of Natural Resources and Environment & Labour.

It is recognized that reclamation plans need to remain flexible and adaptable to changes in site characteristics, mine plans, geology, and long term end land use changes. Such plans should be practical, achievable and be prepared in conjunction with community input. Specific details may be readjusted consistent with good operations and industry best management practices. Plans should also be evaluated by proponents on a regular basis during operations and updated accordingly. Where guidance provided in this document differs from regulated requirements within the province, regulations shall prevail.

PRINCIPLES OF RECLAMATION

The overall objective of a reclamation plan is to produce a landscape that is safe, stable and compatible with the surrounding landscape and final land use. While recognizing that each mine site has unique characteristics, there are principles that are applicable to all surface mining operations and serve as the foundation for planning and implementing reclamation plans for surface coal mines in Nova Scotia.

Best Practices in Reclamation Planning and Management

Mining operations are temporary land use activities and should be conducted with understanding and respect for the environment. Use of reclamation planning and environmental management that aims for sustainability is encouraged in all aspects of reclamation planning, design and implementation. Plans must be science-based, comprehensive in scope and mitigate against safety hazards and environmental effects. Reclamation should be conducted as the operation proceeds.

Applied Principles of Ecological Restoration

A central purpose in reclamation planning should be to promote the ecological integrity of each site and surrounding landscapes. The application of ecological restoration principles requires that plans are developed consistent with regional or landscape level ecological objectives. At the local scale, this involves an examination of surrounding landscapes, in combination with determining predicted successional trends of vegetation communities appropriate to enhance local and regional ecosystems. At the site level, emphasis is placed on reclamation techniques such as land-form replication and planting species that will promote site stability and sustainability. Revegetation should use native species that contribute most to the compatibility of the local ecology.

Compatibility in Land Use, Land Cover and Landscape Design

Surface mining also has the potential to visually impact natural scenery, open landscape character and/or the cultural landscape of adjacent lands. Final reclamation plans and designs should ensure that post-mine or sequential land use or land cover objectives are identified, clearly described and compatible with surrounding landscapes.

Landscape design and visual impact assessments should be incorporated into the reclamation planning process. After reclamation is completed and the operation closed, the site should be self sustaining and/or suitable for an identified or predetermined future land use.

Public Consultation and Informed Decision-making

While respecting landowner prerogatives, surface coal mining activities should provide consideration to community priorities, needs and interests. Reclamation projects can provide lasting benefits to local communities and interested stakeholders can provide important information for plans and decisions that determine reclamation objectives and final land use decisions. Communication and consultation among all parties should be comprehensive, meaningful and timely. Consultation tools can and should include citizen liaison committees and public information sessions.

STANDARD SUBMISSION REQUIREMENTS FOR A RECLAMATION PLAN

Proposed reclamation plans will include the following requirements:

- o proposed life-span of the mine
- o schedule for progressive mine reclamation
- o time-line for the completion of final reclamation and closure
- o existing land use
- o range of post mine land uses evaluated

- o technical and socio-economic factors associated with the range of post-mine land uses evaluated
- o the proposed post-mine land use (natural, conservation or redevelopment)
- o existing features of environmental significance
- o description of any planned biodiversity offsets (if applicable)
- o scaled figures and maps depicting the projected maximum footprint of the proposed mining area over the life of the mine
- o detailed plans for progressive mine reclamation
- o a conceptual final reclamation plan for the final closure of the mine
- o a mitigation and monitoring plan for subsidence, spontaneous combustion and acid rock drainage where appropriate
- o summary of any public issues raised in relation to mine reclamation and how they have been addressed
- o a detailed cost estimate associated with reclamation

PLANNING FOR RECLAMATION

A clear objective for reclamation and final land use should be defined by mine proponents and/or the land owner. Taking the time to plan the various aspects of reclamation early in a project will be of benefit. Reclamation planning should reflect the four principles of reclamation and consider, but not be limited to, the following elements:

- o *Community Involvement*: Extraction is a temporary activity and as temporary occupiers of the land, mining proponents should conduct their business to facilitate post-extraction land use. In proposals for re-development, community and environmental stewardship should be included in the planning and operational stages of the plan. In either case, proponents are encouraged to engage the community during the planning process.

- o *Progressive Reclamation*: Proponents are encouraged to use progressive reclamation whenever possible.
- o *Visual Impact Assessment*: Surface mining has the potential to visually impact the natural landscape. A visual landscape assessment incorporated into reclamation planning should provide certainty that the final site design will be compatible with viewsheds within surrounding natural landscapes.
- o *Compatible Land-use / land Cover*: Reclamation and mine closure should provide land that is restored to a condition that matches surrounding land cover or accommodates another land use identified in the final reclamation plan. Subsequent uses can be oriented toward conservation, recreation or development objectives or combinations of these. Final reclamation plans should ensure that subsequent land use/land cover objectives are clearly identified, described and are compatible with the surrounding land use and landscape.
- o *Top Soil Management*: Sustainable practices applied to soil management and replacement are vital to establishing a self-sustaining cover of vegetation in reclaimed areas. Top soil should be preserved for reclamation wherever possible and soil quality should be protected during moving and storage. Soil management should consider issues such as:
    - o quality assurance during excavation,
    - o identifying stockpile locations to maintain soil quality
    - o temporary seeding
    - o permanent vegetation of stockpiles to control erosion and invasive plant species

Natural chemical fertilizers should be used in addition to considering moisture-holding capacities, drainage, mulches, and compaction of the soil. Where little or no topsoil exists prior to mining, it may be necessary to amend, manufacture or import soils depending on the final land use and site conditions.

- o *Revegetation Planning*: A key element of successful reclamation projects is the establishment of a self-sustaining, succession-based vegetation cover through the application of native seed mixtures and plant materials. This should include a "revegetation plan" to establish the goals for vegetation at the site at the start. Revegetation plans should mimic surrounding non-disturbed areas or encourage specific ecosystem establishment, incorporating strengths of both native and non-native species and strengths of natural and managed growth. This approach is referred to as successional reclamation, which refers to a multi-staged process that relies on different treatments over a period of time.
- o *Ecological Restoration using Sucessional Revegetation Methods where Appropriate*: This approach to reclamation planning is encouraged so that the eventual plant community promotes native species. Selection of native species is encouraged. Through this process the initial use of non-native species will be succeeded by native species. This cover is important as it is considered to be the bridge between initial colonizers and later developing vegetation. Protection of remaining patches of the original vegetation is encouraged to increase seed propagation and improve conditions for natural regeneration of native species.
- o *Prevention of Erosion and Sedimentation*: Grasses and legumes may be required for temporary erosion control and soil rehabilitation. Reclamation plans that involve erosion protection with revegetation must aim to encourage native vegetation and describe appropriate seed recipes, seeding and transplanting techniques.
- o *Prevention of Acid Rock Drainage*: For sites that have been identified as acid generating, reclamation plans to avoid acid rock drainage must be considered at the start of mine planning and carried forward through final rehabilitation and monitoring.

- o *Site Disturbance*: Issues from past mining operations such as acid rock drainage, old infrastructure, subsidence, underground workings, spontaneous combustion of waste and open portals should be addressed in the reclamation plan. Remediation of historical mine features can be a positive result from surface mining and needs to be identified.

LANDSCAPE DESIGN

Landscape design or re-shaping a site is an essential aspect of the reclamation plan. Good mine planning and design can reduce the amount of re-shaping of waste or overburden piles that is required during or after completion of mining. The final land form must be stable and hydrologically compatible with the surrounding area. The objectives for land forming that should be included in the plan are as follows: o provide stability by reducing slope angles. Slopes greater than 3:1 (180) should be kept shorter than 20 metres by means of breaks in the slope, such as uneven or irregular terraces, berms or basins.

- o minimize erosion. Steeper slopes result in increased surface water runoff velocities. Long, uniform slopes should be avoided as they may erode severely during a single storm event.
- o contour aesthetically pleasing landforms consistent with surrounding topography.
- o promote successional reclamation.

REVEGETATION

Revegetation of disturbed lands from surface coal mining is essential for effective and successful reclamation. Planning revegetation requires a knowledge of the pre-existing conditions and a vision for the long term integrity of the system that remains after mining is completed. Reclamation plans will include specific strategies that are used to:

- o stabilize surface materials and prevent erosion
- o manage topsoil

- o enhance natural vegetation growth and establish self-sustainable vegetation growth using native species
- o ultimately support future land use identified in reclamation plan

When determining appropriate revegetation measures for a site, the following items should be considered:

- o future land use or land cover (either natural, conservation or redevelopment)
- o climatic conditions including mean daily temperature, the frost free period, the growing season, the amount and timing of precipitation and the prevailing wind
- o size of the area requiring revegetation in order to assess material requirements (e.g. seed and soil amendment)
- o contouring to mimic local topography and blend into surrounding landscape consistent with future land use or land cover
- o presence of water bodies, wetlands, sensitive ecosystems or other special considerations both pre and post mining
- o availability of stockpiled materials for revegetation
- o success of natural revegetation and species present contouring or constructing works to ensure proper drainage or re-establish previous drainage
- o presence of erosion prone areas, and the necessity for erosion control work, including the use of bio-engineering techniques.
- o soil characteristics including texture, pH, moisture regime, soluble salts and content of nutrients and organic matter and required amendments that may affect revegetation success
- o use of original or native species present on the site
- o re-use of soils on the site that were shifted during mining activities
- o timing of seeding to coincide with optimal germination times of April 15-June 1 and August 15-October 15

- o timing of seeding to coincide with placing the final lift of soil

SITE PREPARATION

Reclamation plans should address the following :

- o an overburden analysis of each major strata on the proposed mine site to determine rock type, texture and thickness and to analyse for chemical and mineral parameters in order to determine what strata is best suited for the top lift of mine soil for revegetation
- o the identification early in the reclamation planning process where topsoil substitutes from deeper soils on the site are required and to store this material separately
- o the application of top soil to a depth sufficient to maintain adequate root growth and nutrient requirements
- o the incorporation of organic materials, mulches, pH adjustment, fertilizers and any other soil amendments based upon soil assessment
- o the mixing of original topsoil with deeper topsoil substitutes at final grading to ensure proper inoculation of soil microbes and the slow release of nutrients
- o minimize compaction of the final lift of soil by end dumping and placement with dozing equipment to avoid excessive compaction
- o the scarification or ripping of flat surfaces which may have been compacted by heavy equipment
- o providing appropriate site drainage to prevent ponding and water erosion on revegetated and other rehabilitated areas
- o establishing windbreaks to prevent wind erosion on revegetated and other rehabilitated areas
- o the use of annuals/perennials for quick cover and provision of soil nutrients for successional vegetation

When revegetating transportation or utility corridors or other disturbed areas, the following rehabilitation measures should be considered for implementation into the reclamation plan;

- o scarifying or ripping corridors after they are no longer required for site inspection and monitoring, so that vegetation can be established
- o grading and contouring to fit the surrounding landscape
- o applying topsoil and other amendments where required to improve initial and sustainable growth
- o any additional measures as required to promote adequate and sustainable revegetation

SITE MONITORING AND MAINTENANCE

Reclamation plans should include site monitoring, and schedules, by the proponent for the following:

- o monitor progress of revegetated areas following initial planting until vegetation is successfully established
- o regularly conduct soil analyses for nutrients and pH until vegetation is demonstrated to be successfully established
- o take corrective action in areas showing evidence of erosion, sedimentation or slope failure
- o take appropriate measures to address evidence of excessive vegetation stress, invasive species or poorly established areas
- o a longer term monitoring schedule to determine any necessary site restoration repairs, and to review progress toward development of a self sustaining ecosystem

BACKFILLING

Backfilling of surface mines is highly recommended. Otherwise, methods that ensure public and wildlife safety must be identified in the reclamation plan. Backfilling is critical in or near areas populated by humans. If the post-mining land use includes structures located on backfilled excavations, the reclamation plan needs to specify how and when the backfill material is structurally sound and stable.

Considerations related to backfilling include:

- o any contaminated materials such as spilled fuel oil, asphalt and solid waste must be removed from the site and disposed of at an approved facility prior to backfilling
- o construction and demolition debris shall not be used as backfill
- o the prevention of subsidence in backfilled material
- o the estimation of compaction and settling rates and how these will affect future land use

FLOODED EXCAVATION MANAGEMENT

Flooding may be used to rehabilitate an open excavation if justified in the closure plan and such flooding is part of planned future land use. The reclamation plan to support a flooded pit must include:

- o final maximum and minimum water elevations
- o to prevent inadvertent walk off and safety issues, the shoreline should be graded to a gentle slope which extends 2 metres below the low water mark of the flooded pit
- o if pit flooding is planned, the potential for acid mine generation or mitigation must be addressed in the reclamation plan
- o shoreline and riparian revegetation should be established to enhance the development of aquatic or wetland habitat
- o overflow from flooded excavations should be directed to a properly designed drainage ditch such that siltation and erosion are minimized, and remain below levels established in approvals and/or regulations

OTHER OPTIONS

Sloping the final pit walls may also be used to reclaim a surface mine if justified in the closure plan as being more appropriate than backfilling or flooding, and sloping "fits" with planned future site use.

If backfilling, flooding or sloping are not practical, then boulder fencing or berming may be used if fully justified in the closure plan and this option "fits" within the planned future use of the site.

Where a surface mine has a single vertical or near vertical drop of greater than 3 metres and a bench width of less than 3 metres and is not to be reclaimed by sloping or backfilling, a geotechnical study and report signed by a professional engineer shall be provided to state the long term stability of the structure.

DECOMMISSIONING OF EQUIPMENT AND INFRASTRUCTURE

All equipment and buildings erected on site for the mining operation should be dismantled and removed as part of the reclamation process, unless they form part of the future land use infrastructure.

Buildings

Buildings may be constructed to last a few years or several decades, depending on the anticipated life of the mine operation. Buildings for short term use should preferably be prefabricated structures that can be easily dismantled at the time of closure and reclamation. Such structures should be totally removed from the site and may be disposed of at an approved facility or reused elsewhere.

Foundations comprised of concrete should be broken up, buried or removed in accordance with applicable waste management regulations.

Equipment

Equipment may include items such as crushing and screening units and coal wash plants. All units must be removed from the site as part of reclamation. Fixed equipment and structures such as weigh scales may also have concrete pads or footings and these should be broken up, buried on site or removed in accordance with applicable waste management regulations.

Power Lines

If surface access is necessary during construction, efforts should be taken to minimize disturbance. During reclamation, all power lines, cables, towers and guy wires should be removed as well as any concrete footings or slabs where appropriate. If access roads were necessary for construction or decommissioning of power lines, they should be ripped and revegetated.

Roads

Access and on-site roads should be properly designed and constructed as part of the initial operating plan to minimize adverse environmental effects and facilitate reclamation. In the operating plan, roads should fit the topography to minimize unnecessary earth moving for road cuttings and embankments. All culverts and drains should be removed and original drainage restored as much as possible in accordance with future land use planning.

WATER RETENTION AND TREATMENT PONDS

During progressive and final reclamation, ponds and other water control and retention devices must be designed to retain all sources of flows (including flows from pumping and surface runoff). As a minimum, a 1 in 25 year storm event should be considered. Water control and retention devices must provide treatment as required for suspended solids removal, and general water chemistry, to comply with baseline local background water conditions and/or regulatory limits prior to water release from the site. Discharges must be designed to operate under controlled conditions, and the plan must reflect this.

Flocculent can be used to enhance removal of suspended solids. Plans that propose the use of flocculent will specify the type and toxicological properties of the flocculent. Water control structures (e.g. spillways, weirs) must be designed to match anticipated life span of the ponds.

Note: The design of water retention ponds/dams must be planned to eliminate the need for water treatment and structural maintenance in perpetuity. The plan should include the most probable duration that ponds or dams will be required and a maintenance schedule that the proponent willundertake for these structures.

ACID ROCK DRAINAGE CONTROL AND MONITORING

Coal mine drainage ranges in composition from acidic to alkaline. Acid rock drainage (ARD) from mines can be formed by the oxidation of pyrite. Mine reclamation practices can influence post-mining water quality, including the extent to which ARD is generated. Rock geochemistry and lithology determine post-mining drainage chemistry. Therefore, it is essential that an initial operating plan contains an assessment of the potential for acid rock drainage from the site. Also, the reclamation plan must be designed to prevent or minimize formation of acid drainage post-mining. Implementation of effective prevention, mitigation and monitoring strategies post-mining must be integral to the reclamation plan where there is a likelihood of acid drainage occurring.

From the standpoint of preventing ARD, the use of vegetation, cover and capping with impermeable surfaces are beneficial for reducing the amount of water and atmospheric oxygen entering the mine soil/spoil environment and should be considered. The grading and revegetation of the surface can also have a substantial impact on the quantity of ARD generated. Prior to revegetation, a layer of non-acid generating material should be placed as a cover over potentially acid generating areas. Amendment of the fill material with limestone to provide additional alkalinity should be considered. A layer of nutrient-rich topsoil of appropriate thickness should then be placed to promote vegetation growth.

MONITORING

Reclamation plans will include a monitoring program with a schedule. Plans shall include, but not be limited to the following elements:

- o a plan to monitor and report on the progress of reclamation throughout mining and post-mining phases
- o a plan to monitor and report on the effectiveness of erosion and sediment control after reclamation
- o a plan to monitor and report on the status of the future land use identified; post-reclamation
- o a plan to monitor the physical stability of the site, to include inspection by the proponent of the following, as a minimum:
  - * slope stability, including signs of new or ongoing slope failure
  - * water levels in open pits for planned minimum and maximum water elevations
  - * the physical stability of all stockpiles including rock, waste, or other mine development stockpiles
  - * erosion, stability and health of vegetation
  - * the physical stability of all water management structures, including ditches and spillways
  - * blockage or potential blockage caused by sediment, ice, debris accumulation or animal activity at water management structures, and
  - * the deterioration of materials used in reclamation to manage site conditions

Note: The frequency of monitoring planned must be adequate to identify issues and be sufficient to ensure public safety and environmental integrity.

- o a plan to monitor and report on potential ARD post mining. Reclamation monitoring for potential ARD should include an appropriate monitoring schedule to ensure that the cover is permanent and will afford the necessary buffering. Vegetative cover assessments should be conducted annually in July or August and should include photographic documentation as well as species assessment, distribution, type, diversity etc.

- o a surface and groundwater monitoring program to be designed by a professional hydrologist or hydrogeologist

WATERCOURSE MANAGEMENT

Watercourses, waterbodies and wetlands must be identified on a site plan for reclamation planning. Watercourse management planning will follow this hierarchy, post-reclamation:

- o ensure that reclamation does not impair water quality in existing watercourses
- o any unavoidable relocation of watercourses and waterbodies must be planned in such a way that returns watercourses to their original location post mining, unless otherwise approved during an environmental assessment or industrial approval
- o watercourse planning and restoration can be used provided those water features fit with intended future land use (ie. creation of ponds, aquatic/riparian habitat restoration or other water related features and instream structures)
- o reclamation planning should incorporate the implementation of an environmental effects monitoring (EEM) program for fish and fish habitat to monitor such parameters as water quality, water quantity, fish and benthic organisms if deemed necessary by NS Environment & Labour

PUBLIC SAFETY

The proponent shall ensure that public safety is of paramount concern and is endorsed within the reclamation plan.

2

Tree Improvement in Northern India

INTRODUCTION

The objective of this chapter is to discuss the current status of Populus tree improvement research, development, and use in northern India. Information was obtained by the senior author from interviews, field trips, and reference materials during two visits to India in 1996. The visits were part of a consultancy activity between Winrock International Institute for Agricultural Development and the Indian Council of Forestry, Research and Education (ICFRE).

USE OF POPLARS IN INDIA

Poplars are primarily used in agroforestry systems on the Indo-gangetic plains north of 28 degrees N latitude (Figure 1) and, to a lesser extent, in forestry plantations on the hills of north India. There is potential for their use in the valleys of the northwestern Himalayas, in the plains and hills of northeast India, and (possibly) for portions of east-central India north of 22 degrees N latitude (states of Madhya Pradesh [=M.P.], Bihar, Orissa, and West Bengal). The crop mixtures used in the agroforestry systems are described later in this paper. Populus deltoides Bartr. ex Marsh. var. deltoides (typical eastern cottonwood from the USA) is an exotic and is currently the species of choice for agroforestry in the Indo-gangetic plains of the states of Uttar Pradesh (U.P.), Haryana, and Punjab. It is also the candidate species for use in east-central India, and may eventually be an important component of poplar culture in northeast India. Indigenous Indian species of Populus do not tolerate the heat of the plains (the possible exception is Populus euphratica Oliv., but this species grows slowly and has a bushy form). Little is known about Populus gamblei Dode in northeast India, but Dr. H. B. Naithani at the Forest Research Institute (FRI) in Dehradun, U.P., does not think that it can tolerate the high temperatures either.

State abbreviations are J&K for Jammu and Kasmir, HP for Himachal Pradesh, P for Punjab, H for Haryana, and UP for Uttar Pradesh. Latitude 28 degrees North has traditionally been the southern limit for use of Poplars in India, but use of subtropical Poplar species may allow extension of this limit to 22 degrees North latitude.

Exhibit heterotic growth in the nursery of the U.P. Forestry Department at Lalkuan, U.P., but fail (dieback) in agroforestry plantings on the U.P. plains by age three (personal communication with scientists at WIMCO Seedlings Ltd. near Lalkuan). A very robust market has developed for poplars grown in agroforestry systems on the Indo-gangetic plains. The primary products are plywood, veneer, and matches, but secondary products include charcoal, fuelwood, leaf litter (composted), and (occasionally) fodder. Farmers are making money from these trees, and many now consider wood as their primary farm crop.

The indigenous poplars, hybrids of P. deltoides with indigenous poplars, and other exotic poplar species and their hybrids with indigenous poplars are proving most suitable for reforestation/ afforestation in the hills of northern India. Dr. D. K. Khurana at the Y. S. Parmar University of Horticulture and Forestry in Solan, Himachal Pradesh (H.P.), recommends P. deltoides clones in the valleys below 900m elevation, the hybrid of P. ciliata x P. maximowiczii Henry on hills between 900m and 1500m elevation, and P. ciliata on hill sites between 1500m and 2800m (personal communication). He also reports that tests are being conducted further north in the state of Jammu and Kashmir (J&K) at Shere Kashmir University of

Agricultural Sciences and Technology on the exotic species Populus nigra L. and P. deltoides, the indigenous species Populus alba L. and Populus sauveolens (?), and the hybrid of P. deltoides x P. ciliata (from his work). D. V. Negi with ICFRE at the Himachal Forest Research Institute (HFRI) in Shimla, H.P., reports that P. alba and P. euphratica only occur in limited, cold areas (high elevations) and might have potential for forest plantations above the P. ciliata plantings or for irrigated plantings in the cold desert region of H.P. and J&K. The U.P. Forestry Department is producing and testing some hybrids of P. deltoides x P. ciliata and P. deltoides with Populus yunnanensis Dode for use in the hills.

Planting with poplars in the hills will be done almost entirely by State Forestry Departments on government lands without benefits of irrigation. Production will not be as rapid as for poplars in the agroforestry systems. Primary objectives shift from wood production to establishing a rapid tree cover for water and soil conservation. However, there is a local market for packing cases to serve the horticultural industry in H.P., so that poplar plantations on state lands might better serve this market than the imported eucalyptus cases from the plains. As the plantations mature, the plywood industry could expand into the hills to utilize the silvicultural cuttings from these plantations.

At least 440 clones of various poplar species and hybrids were introduced and tested in northern India from 1958 to 1983 (Tiwari 1993). There were also some P. deltoides clones sent to India in the mid 1980's by Dr. Sam Foster (then in Louisiana and now with the U.S. Forest Service in Asheville, North Carolina), who made controlled crosses among some selected clones from Stoneville, Mississippi. These clones were tested by WIMCO Seedlings Ltd., but they did not perform well at early ages and were discarded. An additional large collection of approximately 200 P. deltoides clones from throughout the eastern United States was sent by Dr. E. A. Hansen of the U. S. Forest Service at St. Paul, MN, (now retired) to Dr. A. N. Chaturvedi of the U.P. Department of Forestry in January 1986.

Results indicate that P. deltoides from Texas and southern Mississippi River sources are best for the plains of northern India. These origins represent the southern 25% of the native range of P. deltoides, but clones have not been tested from origins east of the Mississippi River in the southern part of the range. P. maximowiczii and P. yunnanensis are other promising exotic species (in the form of hybrids with P. deltoides or P. ciliata) for use in the mid-elevation range of the hills.

In 1990, Dr. D. K. Khurana at the Y. S. Parmar University of Horticulture and Forestry in Solan collected open-pollinated seeds from 103 P. deltoides trees along the Colorado and Brazos Rivers in Texas and along the southern part of the Mississippi River from Baton Rouge, Louisiana, to Davenport, Tennessee (Farmer and Khurana 1990, and personal communication with Dr. Khurana). He raised 50,000 seedlings at Solan and screened them down to 50 clones after three years of nursery measurements. Screenings were based on growth, stem form, and resistance to Melampsora spp. of leaf rust and to stem-borer insects. These 50 clones have been distributed to five state agricultural universities (in Punjab, Haryana, J&K, Orissa, and M.P.), two institutes (Tropical Forest Research Institute [TFRI] at Jabalpur, M.P., and Tata Energy Research Institute in New Delhi), and one private company (WIMCO Seedlings Ltd. in Rudrapur, U.P.). The oldest field trial (at Solan) was three years old in 1996.

TREE IMPROVEMENT–PROVENANCE INFORMATION ON INDIGENOUS POPLARS

There are six indigenous poplar species identified by Tiwari (1993) in India: P. ciliata, P. alba, P. euphratica, P. gamblei, P. jaquemontiana var. glauca (?), and P. laurifolia Ledeb. Dr. D. K. Khurana at Solan also identifies a seventh species: Populus sauveolens (personal communication). These species occur in the hills of northern India, usually as scattered individual trees or as a few widely separated stands. The only provenance studies conducted to date have been for P. ciliata, P. alba, and P. euphratica, and almost all of the reported work has focused on P. ciliata in H.P. Drs. P.K. Khosla and D.K. Khurana (both at the Y.S. Parmar University of Horticulture and Forestry in Solan) have initiated numerous studies on variation in Populus ciliata.

They identified four ecological blocks ["types"] of natural occurrences of P. ciliata: (1) High-level association with fir-spruce, (2) Low-level association with blue pine (Pinus wallichiana A.B. Jacks) and deodar cedar (Cedrus deodara Roxb. Loud.), (3) Ravine/seasonal-water-source association with Alnus, Ulmus, and Salix, and (4) River flood basin/water channel sites on sandy soils (Khurana and Khosla 1982 and 1991). Phenotypic variation is greater among trees within stands than between stands within a type, but large differences are found between types. Provenance tests of P. ciliata that have been reported in the literature (Chauhan and Khurana 1992; Chaukiyal et al. 1995) involve cuttings from one clone (ortet) per source and from only nursery observations at one nursery near Solan. There is some evidence from these tests that better growth occurs from U.P. sources, but fewer limbs and better resistance to stem borers are found from J&K sources.

Recently, Dr. Khurana at Solan has established a larger provenance test of P. ciliata to rectify the problems of too few trees in the earlier tests (personal communication). He collected cuttings from 85 provenances (trees) throughout J&K, H.P., and U.P., and these are now 5 to 7 years old in field trials at three sites in H.P. He has also collected open-pollinated seeds from these same 85 trees, but the seedling clones have not yet been planted in field provenance tests.

D.V. Negi at HFRI in Shimla has recently completed a survey of P. ciliata in H.P. and P. alba and P. euphratica in H.P. and J&K (personal communication). In 1996, he collected and planted cuttings in Shillaru Nursery, H.P., from 124 trees (one male and one female tree in each of 62 stands) for P. ciliata, 32 trees (16 stands) for P. alba, and six trees (three stands) for P. euphratica. None of the P. euphratica cuttings survived. He plans to return to the trees (ortets) of P. alba and P. euphratica to collect open-pollinated seeds. This is part of a rangewide survey proposed to FAO[3] and coordinated by P. Khanna (FRI). Collaborators from FAO are Dr. J. B. Ball,

Senior Forest Officer Plantations, and Dr. Oudara Souvawnavong (from Laos), who are located with the Forest Resource Development Service of FAO in Rome.

Dr. H. B. Naithani (Botany Division, FRI) has been assigned responsibility for surveying P. ciliata in U.P. as part of this project. He has surveyed approximately one-half of the area in U.P. and has taken measurements on 200 trees. He will complete the survey in 1997, but has no plans to collect cuttings or seeds until so instructed by Mr. Khanna. The responsibility for a survey of P. ciliata and P. gamblei in northeast India has been assigned to Dr. Jasbir Singh and Mr. Sharma at the Institute of Rain and Moist Deciduous Forest Research (IRMDFR) in Jorhat, Assam. No accomplishments have been reported from that institute.

TREE IMPROVEMENT–BREEDING

Breeding involves crossing among selected trees to produce progenies with new combinations of genes for testing and selection. The resulting selections will be the clones of the next generation of improvement. Breeding requires flowering trees of the selected clones that will be used as parents. This is a major problem for exotics like P. deltoides, since almost no provisions have been made to establish and maintain breeding orchards of mature individuals of the introduced clones.

All controlled crosses involving P. deltoides before 1997 have been opportunistic, depending on what was flowering. Seven P. deltoides clones flowered during this time (G-48, D-121, and S7C8 were females, and G-3, S7C1, S7C15, and S7C20 were males). Except for D-121, all of these were from Brazos County, Texas. One male flowering clone of P. yunnanensis and one male of P. `robusta' (Euramerican hybrid) were also used. "Breeding" in U.P. has been by the U.P. Forestry Department's Lalkuan Research Centre, Silviculture Division, Sal Region, by nearby WIMCO Seedlings Ltd. in Rudrapur, and at FRI in the Genetics and Tree Propagation Division. Initially, most of the new clones came from open pollinations between G-48 and G-3 (the first two clones to flower).

These seedling progeny clones provided the "L-series" clones of the U.P. Forestry Department and most of the "WSL-series" clones of WIMCO. Subsequently, the U.P. Forestry Department nursery manager at Lalkuan made the following crosses (female x male): G-48 x G-3, D-121 x S7C1, G-48 x P. ciliata, D-121 x P. ciliata, G-48 x P. yunnanensis, and G-48 x P. `robusta'. WIMCO selected three seedling clones from their own controlled crosses of G-48 x G-3, and they made some backcrosses (such as G-48 x P. `robusta) to stabilize desired hybrid combinations of P. deltoides with P. nigra. Controlled crosses by the Noh, E.R., K.K.Sharma, and M.L.Kapoor. 1995. A report submitted to FAO on Improvement Programme of Indigenous Poplars with Particular Reference to India. FRI, Genetics & Tree Propagation Division, Dehradun.

Genetics and Tree Propagation Division at FRI were made in the spring of 1996 and required transport of flowering branches from Lalkuan. The crosses G-48 x G-3 and G-48 x P. ciliata were successfully accomplished. The P. ciliata male flowers came from the hills of nearby Mussoorie.

The greatest amount of controlled crosses among Populus species in India has taken place at the Y. S. Parmar University of Horticulture and Forestry in Solan, H.P. Flowers (male and female) of P. ciliata could only be successfully ripened on cuttings in the greenhouse if the cuttings were cleft-grafted to potted rootstock seedlings in January before bud break (Khurana 1989, Khurana and Thakur 1995). Water cultures and 'twig in pot' methods did not work. Cuttings of P. gamblei bearing male flowers could not be forced to ripen and shed pollen at Solan under any method, so no crosses with this species were made. Dr. Khurana successfully produced the hybrids P. ciliata x P. maximowiczii (pollen obtained from Japan), P. ciliata x P. yunnanensis, and P. deltoides (female) x P. ciliata (male) (but not the reciprocal) (personal communication).

The P. ciliata x P. maximowiczii hybrid was heterotic for growth, but had the branch knot problem of P. maximowiczii. Dr. Khurana selected 20 clones from field tests of this hybrid for minimum branch knots.The P. ciliata x P. yunnanensis and

P. deltoides x P. ciliata hybrids did not show heterosis for growth in his tests. Backcrosses of the Euramerican hybrid clone '1-455' (female), which exhibits fast growth but is highly susceptible to Melampsora spp. of leaf rust, with P. deltoides (male) by Dr. Khurana produced two selections that exhibited extremely fast growth (4 to 5 meters per year in height) on recently-exposed, well-drained forest sites (not agroforestry) in H.P.

A total of 358 clones (primarily P. deltoides, including some intraspecific crosses of G-48 by G-3) have been preserved in a germplasm bank at FRI in Dehradun, U.P. There are also 277 clones of P. deltoides in a germplasm bank at the HFRI Shilly Research Nursery in Solan, H.P.

Most of these are duplications of the clones at FRI, so that protection against a catastrophic loss of clones (from fire, etc.) is provided. The U.P. Forest Department's Lalkuan Research Centre is also maintaining a collection of clones from previous introductions and from inter-and intraspecific crosses. Many of these are duplications of clones at FRI. There are 150 clones in the Department's nursery at Lalkuan (100 P. deltoides clones), another 60 clones in a replicated test planted in February 1985 at `Ganga Pur Patia East' near Lalkuan, and 233 clones in the Department's populetum planted in February 1989 at Tanda, U.P. (Plot #47). Forty-four of the clones in the Tanda planting came from Dr. Hansen's 1986 shipment of 200 clones to India and were selected based on nursery performance at the Lalkuan nursery. The Ganga Pur and Tanda tests were just beginning to flower in 1996 and will provide the only breeding orchard available in India (108 clones) for the initiation of an advanced-generation mating design with P. deltoides.

NURSERY AND GREENHOUSE TECHNIQUES

All Populus propagules being commercially planted in agroforestry or in reforestation/ afforestation are E.T.P.s (Entire Transplants). These are one-year-old rooted cuttings. The tree improvement program must therefore use E.T.P.s for clonal field trials.

Nursery management techniques include soaking the cuttings and nursery beds in water before planting and then planting the 18-25cm length cuttings at a spacing of 80cm x 60cm. Debudding to remove young limbs on the E.T.P.s is done from June to November. The nursery beds are flood irrigated 1-2 times per week during March-June, until the rainy season begins. Backpack sprayers are used to apply insecticides for control of leaf beetle, thrips, and wood-borer insects.

The target E.T.P. for production nurseries is a plant whose ground-line diameter is 1/100th of the height (in cm) (personal communication with Dr. J. P. Chandra). When harvesting the one-year-old E.T.P.s in December-February, the tap root is cut at a 25cm depth and all side roots more than 10cm long are trimmed (Chaturvedi 1982, Sidha et al. 1990). The plants are packed, transported, sold, and planted with a naked root.

Research nurseries have the additional task of developing uniformity among E.T.P.s of the same clone. Re-propagating E.T.P.s annually from the previous year's E.T.P.s for 4-5 years in the nursery may be required to remove "C-effects" (age, position-in-tree, or vigor effects on cuttings taken from the crowns of different-aged ortets on different sites). The clones will not be taken to the field trials until within-clone variation is less than 10-15% (personal communication with Dr. Khurana at Solan). Furthermore, the multiplication of seedling-derived clones will take at least three years in the nursery to produce enough E.T.P.s for field trials at 2-3 sites. When seeds are produced from controlled crosses during breeding, they must be germinated and grown before they can be vegetatively multiplied as E.T.P.s. P. ciliata seeds are very fragile and will lose viability in 7-10 days if not germinated quickly (personal communication with Mr. D. V. Negi).

All Populus species seedlings are vulnerable to sun scald (heat) damage and diseases (damping off) during germination. Procedures for germinating and growing seedlings at WIMCO Seedlings Ltd. provide a working model for success. Fungicide-treated seeds are planted in rows on moist, heat-sterilized sand in flat clay pots and placed in a double-walled polygreenhouse with fans. The polygreenhouse is covered with 50% shade cloth.

Approximately 15 days after the seeds have germinated, the germinants are 'pricked' and transplanted to individual cells in container racks. The soil in the containers is a barnyard mixture of manure that has been sterilized, so no additional fertilization is required. Containerized transplants are placed in a mist chamber in the greenhouse for one week (mist for 1-2 seconds at 5-minute intervals), then moved into a shadehouse for one week and subsequently outside under the shade of trees for 15 days.

Finally, the acclimated seedlings are placed in the open sunlight where they remain until they are 30cm tall. At this size the roots can retain soil in a 'plug' when removed from the container. The 'plug' seedlings are then planted in the research nursery (personal communication with Dr. S. Chauhan of WIMCO).

AGROFORESTRY

The entire agroforestry industry with Populus is founded on a very limited genetic base.

Approximately 90% of poplars being planted for agroforestry in U.P., Haryana, and Punjab come from the P. deltoides clones G-48, G-3, and S7C15 (personal communication with Dr. J. P. Chandra of WIMCO Seedlings Ltd.). The next two clones in popularity are 'Udai' from WIMCO and L-34 from the U.P. Forestry Department. Both of these clones come from open pollinations between G-48 and G-3. Other clones that are not quite as desirable, but which are being held in reserve in case of a serious disease or insect outbreak, are S7C8, S7C4, L-43, ST-240, ST-70, and ST-67. All of these clones (except the three ST clones) came from gene pools in Brazos County, Texas. The three ST clones came from the southern Mississippi River alluvial plain just north of Vicksburg, Mississippi USA.

Typical agroforestry methods are to plant 4m-5m tall E.T.P.s in (1) "bund" (shelterbelt) plantings on the small levees around flood-irrigated agricultural fields, (2) "block" plantings at 5m x 5m, 5m x 4m, or 6m x 4m spacings within agricultural fields (underplanted with the crops), or (3) "row" plantings that contain alternating rows of Poplars and horticultural tree species underplanted with agricultural crops. Typical spacing between "rows" is 6m, so the P. deltoides rows are 12m apart. Rotation lengths for poplars are 5-8 years (5 years for plywood, 8 years preferred for veneer and matches). Typical crop combinations with "bund" plantings are sugarcane or rice, but rice is not recommended because of the requirement for summer flooding.

Crop combinations with "block" plantings involve sugarcane for the first two years and then winter wheat or a combination of winter wheat alternating with summer-grown tumeric or pearl millet during the third through eighth years of the P. deltoides rotation. A new alternative to tumeric or pearl millet is celery. Dr. N. B. Singh has also observed rice being cultivated in some block plantings during the fourth through eighth years. Apparently, P. deltoides can tolerate flooding at the older ages. "Row" planting mixtures in agro-horti-forestry systems are based on a rotation length of up to 50 years for the final horticultural orchard. These systems may involve poplar-peach, poplar-litchi, or poplar-mango combinations with sugarcane, tumeric, and winter wheat as under crops. Up to two 5-to 8-year rotations of Populus may be grown before the fruit trees (particularly mango) spread in crown diameter to become a full orchard. In all of these systems the poplar trees are pruned of lower limbs (recommended lower 1/3 of stem height after the second year). The leaves are collected and composted at the end of the growing season for placement back on the fields, and the stumps are dug up and removed for fuelwood/charcoal at the end of the rotation.

SUMMARY COMMENTS AND RECENT DEVELOPMENTS

There has been much plant material, expertise, and work involved in Populus tree improvement for India. However, there has also been little coordination of effort or continuity of guidance, since long-term assignments to projects have not been common. Although the widespread testing of many introduced clones, release of a few selected clones, and development of a new income source for farmers is worthy of praise, much more should have been accomplished in 40 years.

Dr. Khurana in Solan has been very astute in seeing and working on the tasks that were most needed:

(1) Breeding of superior clones of P. deltoides and a few other exotic poplar species that were already present and proven in northern India (i.e. "land races"),

(2) Provenance testing of indigenous poplars for performance and for amounts and patterns of geographic genetic variation,

(3) Introduction of a larger gene pool (through seed collections) from the portion of the natural range of P. deltoides that has proven to be most productive in northern India, and

(4) Widespread testing of new clones from breeding, provenance testing, and new introductions across all of northern and east-central India.

The Indian government has recently recognized the need for central coordination of forestry research and development activities and has assigned this task to the Indian Council of Forestry, Research and Education (ICFRE). Desirable steps have already been taken. In mid 1996 the Director General of ICFRE appointed Dr. N. B. Singh to be Chief Technical Advisor (Coordinator) and Mr. Dinesh Kumar to be Associate Technical Advisor for an All-India Coordinated Project on Poplar Improvement (AICPPI). These two scientists are located at FRI in Dehradun, which is where ICFRE administration is headquartered. Priorities will be:

(1) Advanced-generation breeding of land-race' clones of P. deltoides and introductions of new collections of P. deltoides from the southeastern USA for agroforestry in the alluvial plains of northern India (above 28 degrees N latitude),

(2) Developing indigenous poplars (P. ciliata and P. gamblei) and their hybrids with P. yunnanensis and P. maximowiczii for the hills of northern India, and

(3) Initiating introduction trials of subtropical and tropical poplars (P. ilicifolia Rouleau from the Tana River of

tropical Kenya and P. mexicana subsp. mexicana, P. mexicana subsp. dimorpha, P. fremontii subsp. mesetae, and P. guzmantlensis from subtropical Mexico) (Pryor 1992) for agroforestry between 22 and 28 degrees N latitude in central India.

Several activities have already been accomplished at FRI during December 1996 through March 1997. A National Poplar Germplasm Bank with 350 clones has been established, 45 promising clones of P. deltoides have been planted in a breeding orchard, control crosses of P. deltoides have been made among six female clones and 15 males, and cuttings of 20 P. deltoides clones have been supplied to 15 cooperating research institutions for multilocational trials. Perhaps with the new AICPPI the long-term commitment of resources needed for a Populus tree improvement program can be sustained. Such an approach is necessary to efficiently achieve the main goal of continually providing better planting material and products for the nation.

3

Eucalyptus Tereticornis and Native Dalbergia Sissoo Tree Plantations

INTRODUCTION

Pre-1988 Forest Policies in India promoted wide-scale plantations of exotic tree species such as Populus deltoides, Acacia spp., Eucalyptus spp., Leucaena leucocephala and Prosopis juliflora to meet the increasing industrial and fuel-wood demands of the public (Bajaj 1997). Eucalyptus tereticornis was preferred to other exotic trees because of short-term visible gains for straight bole, fast growth rate, more productivity per unit area and least post-plantation care (Mathur et al. 1984; Kushalappa 1985; Rajan 1987; Sharma et al. 1988; Kapur and Dogra 1989; Chatha et al. 1991). But the scientific community, private growers and the public have been divided over the merits and demerits of Eucalyptus plantations for economic gains from wood, and for ecological functions such as water usage, understorey ground cover and allelopathic effects (Mathur and Soni 1983; Dabral and Raturi 1985; Shiva and Bandyopadhyay 1987; Bahuguna et al. 1990; Narain et al. 1990; Geetha et al. 1994; Jalota and Kohli 1996; Jalota 1997; Jalota et al. 2000; Sangha et al. 2000; Singh and Singh 2003).

The total value estimations based on monetary returns from wood in short rotation exotic plantations led to undervalue native trees such as Dalbergia sissoo, mainly due to their longer life span. Many native trees, e.g. Dalbergia have better timber quality than Eucalyptus (Shiva and Bandyopadhyay 1987) and are preferentially used for good quality furniture.

The ecological services (shade, shelter, fodder and medicinal value) rendered by native trees are essential for the common man, especially in some rural communities of India, but the value of these 'unseen' ecological benefits is invariably ignored in our accounting system for forest resources. Moreover, the negative effects of exotic plantations, such as reduced plant diversity (Jalota 1997) are neglected in such an accounting system. Hence, they fail to provide an accurate estimate of costs and benefits of native or exotic plantations.

Most of the reports available to date (Bahuguna et al. 1990; Chatha et al. 1991; Kushalappa 1985; Mathur et al. 1983; Rajan 1987; Shiva and Bandyopadhyay 1987) have focused on either economic benefits from wood products or on the ecological aspects of exotic and native tree plantations, but none has integrated ecological and economic potential for tangible and intangible benefits. Jalota and Sangha (2000) compared the tangible benefits of Dalbergia and Eucalyptus plantations, and showed that over a short-term the tangible benefits from Eucalyptus were greater than Dalbergia, while these benefits were greater for Dalbergia if considered over 21 years of age. This led us to consider the importance of ecological services that are otherwise overlooked such as phytodiversity, nutrient return from litter and soil nutrients, and to rationally evaluate their ecological and economic potential in Eucalyptus and Dalbergia monoculture plantations for two age groups, i.e. young (6-8 year) and old (19-21 year). The monetary equivalence of ecological services is reported here in addition to tangible gains, to determine the total value of Eucalyptus and Dalbergia plantations from the ecological and economic perspectives.

METHODS

Two sets of Eucalyptus and Dalbergia plantations, each 6-8 year (young) and 19-21 year (old), in triplicate, were selected for the study during 1996-1999, in the territory of Chandigarh, in north-west India (30° 42' N, 76° 54' E 333) under similar edapho-climatic conditions.

The selection of these age groups was important as 6-8 years and 19-21 years represent the mature age in Eucalyptus and Dalbergia respectively. A representative area of 4 ha was marked at each site for various measurements.

Ground Vegetation

Species diversity and biomass productivity of various plants growing on the floor of Eucalyptus and Dalbergia plantations were measured during summer, autumn, winter and spring in three consecutive years following the quadrat method (Misra 1968). Ten 1 m x 1 m quadrats were laid randomly in each season at each of the sites. Indices of diversity, dominance, richness and evenness were computed using the statistical software ECOSTATS. The importance value index (IVI) (sum of relative values of density, frequency and dominance) was also calculated for each species. The formulae used to calculate density, frequency and dominance (Misra 1968) are as follows:

- Density of a species = number of plants of a certain species/area
- Relative density = density of a species/total density of all species x 100
- Frequency = number of quadrats of occurrence of a species/total quadrats sampled
- Relative frequency = frequency of a species/total frequency of all species x 100
- Dominance = basal area of a species/total area sampled
- Relative dominance = basal area of a species/basal area of all the species x 100

The average value of the importance value index and the relative biomass of each species was calculated from the seasonal readings taken over a year.

The ground floor species contribute to the ecological sustainability of an ecosystem by contributing to soil stability, maintaining hydrological and nutrient balance. Some of these species are also used in food preparation. A review of the literature (Singh et al. 1983; Kamal 1988; Husain et al.,1992; Rastogi and Mehrotra 1993; CSIR 1994) and local surveys (conducted by the authors) revealed that the plant species found on the floors of Eucalyptus and Dalbergia had five main uses: food, fuel, fodder, medicine and soil stabilisation.

In the absence of local markets to capture the value of these usages, we opted for the ordinal analysis method (Henderson and Quandt 1980) for assigning monetary value to each component of ground-floor vegetation. The monetary equivalence for each plant species was estimated in relation to its use for the selected five categories: food, fuel, fodder, medicine and soil stabilisation. Each species was assigned scores based on the lifespan (longevity) and biomass production. On the basis of longevity, scores were assigned as 1 for annual, 2 for biennial and 3 for perennial life cycles, and for biomass production, scores were assigned according to the plant weight (1 for weight <10 g/plant, 2 for 10-100 g/plant and 3 for >100 g/plant).

The total scores for each species for all five categories were calculated in relation to their life cycle and biomass production as number of uses x (life cycle score + biomass score). For example, a plant with three uses (food, medicine and soil stabilisation) having a perennial life cycle and biomass > 100 g, was scored 18. In the absence of any information on a particular use of a species, its score was assigned the value zero. The function of soil stabilisation was considered in all species, so its value was counted for all species. A plant with all five uses, perennial life cycle and biomass productivity >100 g/plant, scored the maximum, i.e. 30. The total economic value was then computed for all the usages by applying the following method (Chopra et al. 1997 a and c):

$$Pt = Pi \times Si / St$$

where

Pt = Total economic value for all the uses of a plant species;

Pi = Mean market price of a species per plant;

Si = Score obtained for the various uses of a plant;

St = Total maximum score for all the five uses.

The total monetary value of phytodiversity in Eucalyptus and Dalbergia plantations was calculated according to the total number of plants per hectare in each plantation. The total number of individuals of a species per hectare per year was calculated from the seasonal data collected over year (as mentioned earlier). The monetary value of each plant type was then multiplied with the number of plants per hectare of a species to estimate the total value per hectare. While the utility and adequacy of this approach need to be confirmed by other studies, it is useful at the small scale (farm) to assess the value of various usages of plant diversity in the absence of any direct market value. A similar approach was used by Belal and Springuel (1996) to estimate the traditional value of particular uses of plants, however they did not calculate the monetary equivalence for a particular use of plant species. The present approach provides an estimate for use and ecosystem function for each plant species. This approach could be particularly useful in developing countries where people make use of wild plants/herbs in their daily diet and rely upon the natural diversity of these plants.

Soil

At each site, six litter-free soil samples were collected at the same time to a depth of 30 cm during summer, autumn, winter and spring for three consecutive years. The samples were air dried and sieved (0.2 mm mesh). The fresh and dry weights were also taken to determine soil moisture content. The samples were bulked for chemical analysis. The available content of phosphorus (Olsen et al. 1954), nitrogen (Kjeltec system I), potassium (flame photometer), organic carbon (Walkey and Black's titration method, Piper 1950) and micronutrients, viz. Cu, Zn, Fe and Mn (extraction with diethylene triamine penta acetic acid, using atomic absorption spectrophotometer) were analysed. The pH and electrical conductivity (EC) were measured with pH and EC meters. Data were analysed using Duncan's multiple range test (DMRT) (Duncan 1955).

The monetary value was assigned to nutrient content available at a site following the Surrogate Valuation Technique (Chopra et al. 1997b). It uses the information based on a marketable good to infer the value of an associated non-marketable good. This technique provides an economic measure to assign value for soil nutrients based upon the market value of fertilisers.

The value of soil nutrients was computed in relation to the market price of chemical fertilisers (see Chivaura-Mususa et al. (2000) and Guo et al. (2001) for similar approaches). Market surveys were conducted for price value of nitrogen, phosphorus, potassium, copper, zinc, iron and manganese. This helped to account for standing value of soil nutrients and assess it in relation to plantation type.

Soil pH, electrical conductivity (EC), moisture content and organic carbon hold no direct or indirect market value, hence these were evaluated on the basis of quantity and for their positive or negative importance in productivity. For example, soils were scored as-1 for pH < 6.0 (acidic), 2 for pH 6-8.5 (medium range, allows to grow most plant types), and-1 for > 8.5 (alkaline). Soil with EC < 0.8 (normal to grow most plants) was assigned 2 scores, with EC 0.8-1.6 (critical for salt sensitive crops) as -1, and for EC 1.6-2.5 (critical for salt-tolerant crops) as -2 scores. Moisture content was scored as 1 for low (0-3.5%) and 2 for > 3.5%. The content of organic carbon was scored as 1 for < 0.4% (low), 2 for 0.4-0.75% (medium), and 3 for > 0.75% (high).

Total soil value index (%) for pH, EC, moisture content and organic carbon (OC) was calculated according to the following formula:

Total Value Index: S score for soil pH, EC, moisture content and OC at a particular site/total score (i.e. 9) x 100.

Litter Production and Nutrient Release

The amount of litter production was quantified at regular three-month intervals for each season in a year from five permanent 1 m x 1 m quadrats. The measurements were taken only at old plantations of Eucalyptus and Dalbergia. Surface litter decomposition following the litter bag technique (Pauley and Little 1998), and annual release of nutrients (N, P and K) were studied. Nutrient loss over a one-year period of decomposition was computed in relation to the amount of litter produced at each site.

For monetary valuation, the release of nutrients from litter was considered as return of nutrients to the soil. The surrogate valuation technique (as applied for soil nutrients) was used to estimate the value of a particular nutrient. The monetary value of nutrient return calculated for 19-21 year old plantations was considered to be the same for 6-8 year old plantations in the absence of data from young plantations.

RESULTS

Ground Vegetation

Ecological Evaluation: The density of plants (no. of plants per hectare) in young Dalbergia plantations was double that of same age Eucalyptus plantations and more than four times that of old Eucalyptus plantations.Young and old Eucalyptus plantations had 17 and 18 ground species of plants respectively, whereas young and old Dalbergia plantations supported 34 and 28 plant species.

The plant species specific to Eucalyptus were few (three in young and six in old) compared to Dalbergia plantations (20 in young and 16 in old). The importance value index was maximum for Cynodon dactylon in young plantations of Dalbergia and Eucalyptus. In old plantations, the importance value index was greatest for Saccharum munja in Eucalyptus and for C.dactylon in Dalbergia plantations.

The greater diversity of plant species on the floor of Dalbergia compared to Eucalyptus was clear from Shannon's index of diversity. The indices of richness and evenness demonstrated uniform distribution of abundance of different plant species in Dalbergia compared to Eucalyptus plantations.

Economic Evaluation

The number of plant species that have use value for food, fuel, fodder, medicine, and soil stabilisation was greater in Dalbergia compared to Eucalyptus plantations, regardless of plantation age. More plant species with annual and perennial life cycles were present in Dalbergia than in Eucalyptus plantations. The total biomass productivity of all plants was twice in younger, and more than three times greater in older plantations of Dalbergia compared to Eucalyptus plantations.

Based on the criteria of biomass productivity and life cycle of a plant, the monetary equivalence was computed for the five selected uses of a species. In 6-8 year old plantations, C. dactylon had the maximum value among all plant species in Eucalyptus and Dalbergia plantations, but had 1.24 times greater 98 / Sangha and Jalota value in Dalbergia than in Eucalyptus. In 19-21 year old plantations, S. munja had the maximum value followed by C. dactylon in Eucalyptus plantations; and Cannabis sativa followed by C. dactylon had maximum values in Dalbergia plantations. The total monetary equivalence for all plant species was 1.83 times and 5.78 times greater in respective young and old plantations of Dalbergia compared to Eucalyptus plantations.

Soil Nutrients

Ecological Evaluation

In both young and old plantations, the content of available N was in the soil was significantly greater in Dalbergia plantations (1332.95 kg/ha and 1774.57 kg/ha respectively) compared to that in respective young and old plantations of Eucalyptus (1112.8 kg/ha and 1520.74 kg/ha). Similarly, the available content of K was significantly greater in Dalbergia plantations (1173.88 kg/ha at young and 703.75 kg/ha at old plantations) compared to that in respective young and old Eucalyptus plantations (935.15 kg/ha at young and 454.8kg/ha at old plantation soils). Available content of Pdid not differ between Dalbergia and Eucalyptus plantations of the same age group, but it was greater in soils of young plantations (18.51 kg/ha in Dalbergia, 17.71 kg/ha in Eucalyptus) compared to old plantations (16.09 kg/ha in Dalbergia, and 15.33 kg/ha in Eucalyptus). Dalbergia soils also had greater content of Fe and Mn compared to soils of Eucalyptus. Zn content did not show any difference at young age plantations, however, there was a greater content at old age in Dalbergia plantations. Only Cu content was greater in Eucalyptus than in Dalbergia soils.

Soil organic carbon was maximum in old Dalbergia plantations (1.95%), and differed only with Eucalyptus at young age (0.69%) after applying DMRT (at 0.05% level). Although soil pH and EC did not differ significantly between Dalbergia and Eucalyptus plantations, soil EC was more than double in Eucalyptus soils (0.87 dS/m in young and 0.89 dS/m in old plantations) than that in Dalbergia soils (0.31 dS/m in young and 0.42 dS/m in old plantations) which could adversely affect plant growth. Soil moisture content was greater in Dalbergia (6.32% at young and 5.88% at old plantation) plantations compared to Eucalyptus (1.86% at young and 3.66% at old) plantations.

Economic Evaluation

The monetary equivalence for available N, K, Fe and Mn was greater in Dalbergia than in Eucalyptus plantations. Eucalyptus had greater monetary equivalence for Cu content than Dalbergia. Soil Zn content did not differ in its value in Eucalyptus and Dalbergia plantations, except old Dalbergia plantation, which had the maximum value. The total monetary equivalence for all the studied soil nutrients was 1.34 times greater in Dalbergia compared to Eucalyptus at both young and old age of plantations . Dalbergia soils had greater score for soil organic carbon and soil moisture than Eucalyptus soils. This led to a greater total soil value index in Dalbergia (100% in both young and old) compared to Eucalyptus soils (50% in young and 75% in old plantations) .

Litter Production and Nutrient Return

The amount of litter produced in a year was 1.44 times greater in Dalbergia plantations (5698 kg/ha/yr) compared to Eucalyptus (3946 kg/ha/yr). The content of N, P and K was greater in litter collected from Dalbergia plantations compared to Eucalyptus plantations, owing to their greater mass and concentration.

Value of Nutrient Return from Litter

The total monetary equivalence for the amount of nutrients released from litter in relation to the total amount of litter produced in Dalbergia plantations was 2.61 times greater (6709 Rs/ha) than that at Eucalyptus (2561 Rs/ha) .

Total Benefits from Phytodiversity, Soil Nutrients and Litter

The ecological benefits represented only 0.09% proportion of total tangible benefits in young Eucalyptus and 0.39% of total tangible benefits in young Dalbergia, and 0.01% of total tangible benefits in both old Eucalyptus and Dalbergia plantations, when counted only once. But, there is a continuum of benefits from these services over the age of a plantation. Therefore, the value of these services over the age of a plantation i.e. for 8 year and 21 year was calculated, which represented 0.27% of tangible benefits in young and 0.07% of tangible benefits in old Eucalyptus; and 1.38% in young Dalbergia and 0.12% of total tangible benefits in old Dalbergia plantations. Overall benefits from wood, non-wood products, plant diversity, soils and litter nutrient return (computed over the age of a plantation) were 1.6 times greater in Eucalyptus than Dalbergia at 8 year age of plantations. These were chiefly from timber in young Eucalyptus while Dalbergiahad no timber value at this age, but Dalbergia supported 1.8 times greater value of ecological services than that of Eucalyptus.

At 19-21 years of growth, Dalbergiasupported 2.7 times more value for its tangible and ecological services than Eucalyptus plantations (Figure 2). Increment in total benefits with age of plantation was more in Dalbergia (from 1,090,284 Rs/ha at young to 15,008,889 Rs/ha at old) than that in Eucalyptus (1,698,686 Rs/ha at young and 5,561,981 Rs/ha at old plantations).

DISCUSSION

Native plantations of Dalbergia supported greater species diversity, soil nutrient content and litter production for more nutrient return than exotic plantations of Eucalyptus. The meagre understorey plant diversity in Eucalyptus plantations has been attributed to allelopathy (Kohli et al. 1990, Kohli and Singh 1991; Reid et al. 1992; Srivastava et al. 1994; Verma and

Totey 1999) and/or the toxic effects of allelochemicals on soil micro-organisms (Dellacasa et al. 1989; Chander et al. 1995). Poor soil status in Eucalyptus plantations was also reported by Jha et al. (1999). Despite such evidence for the negative effects of eucalypts in monoculture plantations, it has been favoured over native trees due to the economic gains for wood (Shiva and Bandyopadhyay 1987; Jalota 1997). The ignorance of the costs associated with the negative effects of Eucalyptus, and the non-accounting of ecological services of native trees in total value assessment, resulted in the prediction of greater economic benefits for eucalypts over native trees.

Why are ecological services, such as species diversity or litter, so important? Tilman et al. (1997) report that diverse plants perform different ecosystem functions which contribute to the sustainability of an ecosystem. Ecosystem functions performed by one species may be complementary to the other species. Niche matching and the probability of presence of one species in promoting the existence of other species, are important mechanisms that determine species diversity in a sustainable ecosystem (Tilman 1999). According to the diversity-productivity hypothesis (Tilman et al. 1996; 1997), greater species diversity supports greater biomass productivity. Such a relationship was evident for plants growing under the canopy of Dalbergia, which showed greater total biomass compared to plants under Eucalyptus plantations, though there could be other factors such as soil nutrients that promote plant growth under the canopy of Dalbergia.

In addition to ecosystem functions of diverse plant species, various wild plants growing under native plantations are also important as medicine and food. The diversity and growth of plants are related to soil nutrient content and nutrient return from litter decomposition in an ecosystem. In Eucalyptus and Dalbergia monoculture plantations, the ground vegetation, soil nutrients and litter, which provide habitat for soil microorganisms, are important for ecosystem sustainability. Ground vegetation is also important to some extent to fulfil the needs of the local people. Therefore, the main ecological benefits from understorey plants in terms of food, fodder, fuel, medicine and soil stabilisation, from soil for major nutrients that support plant growth, and nutrient return from litter are considered in this study. There are many other ecological services e.g. carbon sequestration, aesthetic value of a plantation and the existence value of diverse plant species, which also contribute to the total value of an ecosystem, which are not accounted for in this study.

The study highlights the importance of native trees for their value of ecological services as the benefits were twice greater in Dalbergia than in Eucalyptus plantations, when calculated over the age of a plantation. Although at young age of these plantations, Eucalyptus scored over Dalbergia owing to its prominent timber gains, at 21 years of age, the total returns from Dalbergia were about thrice greater than from Eucalyptus. There was also a greater increase in benefits from 8 to 21 years of age in Dalbergia compared to that in Eucalyptus.

It is important to note that in Eucalyptus (21 year-old plantation), tangible benefits (timber and other non-wood products) were calculated at a rotation of eight years over a 21-year time period (i.e. for 2.5 crops, as Eucalyptus is generally harvested at 8-10 years of age) (Jalota and Sangha 2000). Our earlier study (Jalota and Sangha 2000) reported the main tangible benefits from these two plantations, though at a young age the benefits were greater from Eucalyptus, but at an old age, Dalbergia performed better than Eucalyptus.

The present results further suggest that Dalbergia plantations are more profitable at 21 years of age with greater ecological services from understorey plant diversity, litter and soil nutrient content than Eucalyptus plantations.

Moreover, over the period of growth for Dalbergia plantations, ecological services will be in a continuum to sustain the system over a longer time, whereas in Eucalyptus rotational harvest at every 8-year intervals will disturb the dynamics of ecological services.

The methodology used here to assign monetary value to intangible services would need further improvements but such a tool is appropriate in a developing country such as India where common herbs/plants are used in daily life. The application of standard techniques (Contingent Valuation Method (CVM), Choice Modelling (CM) Method, Contingent Ranking, Travel Cost Method (TCM), and so on) cannot be generalised as they largely depend upon the culture and financial status of a society (Sangha et al. 2000). Other methods based upon ecosystem functions have been proposed. Nunes et al. (2001) estimated the value of biodiversity at four levels, i.e. genes, species, ecosystems and functions, but they missed the social value attached to a particular usage which is important for people in developing countries.

Human interactions with environment are important when considering valuation of ecological services (Daily 1999). In developing countries, common valuation methods (CVM, CM or TCM) may not work due to regional variations and societal differences to use a plant species, to make aesthetic sense of a site or differences in cultural values. There is a lot of integration/cross-linkage of use and non-use benefits from forests, and their interaction with people; and this emphasises the need to develop some indigenous methods to evaluate ecological services. The methodology used in this paper was based on the actual use of plant species.

This paper emphasises the importance of ecological services in total value over the long term for fast-growing exotic and steady-growing indigenous trees in terms of all direct and indirect benefits before implementing any forest policies/ decisions. The total value judgment over a long term could provide a better idea about the potential of a tree plantation. The development of indigenous methods to evaluate the use of indigenous plants by locals will help to assess their accurate value for future policy decisions.

4

Survey of the Status, Distribution and Home Range Requirements

INTRODUCTION

The Punjab urial Ovis vignei punjabiensis is endemic to northern Punjab, Pakistan, and is currently classified as endangered by the IUCN. The distribution of the animal in Pakistan is between the Indus and Jhelum rivers at elevations of 250 -1,500 m primarily in the Kala Chitta and Salt Ranges. The dominant habitat type in the area is dry sub-tropical, semi-evergreen scrub forest. Punjab urial were once present over all northern Punjab Mountains, but in recent decades underwent a severe decline in both range and numbers, disappearing from much of their historic range.

The main reasons for the rapid decline in numbers appear to be poaching, the capture of newborn lambs that are kept as pets, fragmentation of home ranges, competition with domestic livestock and habitat disturbance and destruction associated with increasing agriculture, forestry and mining. In addition, infusion of exotic but compatible mouflon sheep genes may have occurred. The recent construction of the M-2 Motorway between Islamabad and Lahore created a substantial barrier to seasonal migrations and to dispersal.

The main factors affecting the continued persistence of the wild population are inadequate protection against poaching and habitat loss. Initial conservation measures should include, increasing the law enforcement capabilities, to incorporate community participation in the management of urial and reduce the number of competing livestock.

The present study is also as an attempt to collect basic ecological information on the urial in the Salt Range area. The overall aim of the project is to accurately determine the status and distribution of the Punjab urial throughout its current home range in northern Punjab and to determine what its ideal habitat requirements are.

STUDY AREAS

The Salt Range

The name Salt Range, owing to the second largest mineral salt (Sodium chloride) deposits in the world, is given to the hill system situated in the Northern Punjab, in Jhelum, Chakwal, Khushab and Mianwali Districts. First time the name Salt Range was used in 1808 by Elphinston, a British Envoy, when he observed the extraction of salt in the area. The salt deposits of the area were deposited as a result of the evaporation of Tethys Sea and formation of Indus plains from collision of Indian plate with Asian plate resulting from continental drift (King and Vincent, 1993). The Salt Range is an east-west trending thrust front about 175 km long. It forms an impressive scarp, from 250-1520 m in altitude. Sakesar top is the highest point (1524 m). It extends between 32° 41-32° 56 N. and 71° 50 to 74° E.

This range first enters the Chakwal district at its extreme south west corner where the spurs of mount Sakesar descend into the village Lawa. In this part of its course the range keeps mostly to the district of Khushab, but near Khewra, it passes altogether into the Jhelum district where it bifurcates into two distinct ridges, one of them running towards south-east while the other into the east, about 8 km apart from each other. Each of them is made of a number of small roughly parallel ridges. This parallelism is modified by a marked tendency for linked and looped formations. On the south, the range presents a monotonous line of parched and barren slopes, rarely more than 250 m high from the mean sea level, descending abruptly into the valley of the Jhelum River. On the northern side they gradually sink down into the Chakwal plateau (Lindsay, 1923).

Sedimentary rocks and the fossils preserved therein give a complete record of the geological and biological history of the earth. The rock layers in the area have been tilted vertically, even inverted in some places, so that the older, fossil strewn layers now lie on the surface (Shaw, 1989).

The over use of vegetation has accelerated rates of erosion resulting in bare sheet rocks devoid of any soil layers. The rocks are composed of limestone and sandstone or both. At some places infertile red marl is exposed due to similar reasons and the steep geological tilt resulting in frequent slips. The plant cover is poor on sandstone and red marl. The density of vegetation on southern aspects is poor while the northern slopes are comparatively bettered covered with vegetation.

The habitat type prevailing in the area is dry sub-tropical semi-evergreen scrub forest (Roberts, 1991). The dominant plant species are Acacia modesta, Olea ferrugenia, Salvadora alights, Zizyphus nummularia, Dodonea viscosa, Prosopis glandulosa, Justiciar adhatoda, Calotropis procera. Shrubs are sparse with scattered Zizyphus nummularia and, May tenus Rawlins except in some ravines and on the high ridges where Daytona viscosa is prominent and grasses like Cymbopogon jwarancusa, Eleusine compressa, Heteropogon contortus, Aristida adscensionis, Cynodon dactylon and Saccharum species.

Salt Range had a varied and abundant wildlife species in the historic times. Punjab urial, Chinkara, important carnivores, Chukar, See-see, grey and black partridges were in plenty due to nature of vegetation and topography. Which has been over hunted in the past and led to marked reduction in the numbers and restriction of the range of most species. Chinkara is nearly extinct from the Salt Range.

Punjab urial population is also declining day by day. To protect urial population, one National Park. Five wildlife sanctuaries and two game reserves have been established in its distribution range. But this protective system of management has not been able to rise in its population or even stop decline in population. Chhumbi Surla and Jalalpur wildlife sanctuaries and Kalabagh Game reserve (KGR) have comparatively better urial population than other areas. So these sites were selected for detailed study.

Rakh Topi Game Reserve and Boraka Wildlife Sanctuary

Rakh Topi Game Reserve and Boraka Wildlife sanctuary, these two protected areas supports a thin population of urial. Rakh Topi Game Reserve and a part of Boraka Wildlife sanctuary lies in district Kohat. It is comprised upon Tehsil Lachi that is sandwiched between Tehsil Kohat in north and district Karak and Mianwali on south. Towards the east it is separated from district Attock by the mighty Indus, while district Hungu is located in the west.

Kalabagh Game Reserve

Kalabagh game reserve (KGR) established in the early 1930's, is located about 25 km south east of the town of kalabagh, in Jaba masan valley, Mianwali district, punjab province in a small massif that forms the most westerly extension of the salt mountain range. For many generations, this land has been the private property of the Nawab of Kalabagh.

Lehri Nature Park

Lehri Nature Park comprises an area of about 1325 ha of a bit steep and rugged mountains ranging from 250 to 1025 m in altitude above mean sea level and forms the most eastern extension of the Salt Range. The park centered on Lat. 33° 09 .012N. Lon. 73° 35. 931 E. near about 10 Km from Dina town, in the Jhelum District of Punjab. Grand Trunk (GT) Road from Lahore to Islamabad serves as a boundary of the park in the south.

The Northern edge is at Mangla Reservoir, while East and West boundaries are close to the Mangla cantonment and Lehri village respectively. No maps of the 1: 50, 000 topo-cadastradal series were available for the area; the alternatives being A4-sized sketch maps are available from the Wildlife Department.

Park Management

The project area is the property of the Punjab Forest Department and under the settlement rights has been declared as Reserved Forests (Lehri and Judi). The Punjab Forest and Wildlife Departments manage the park. The protection of wildlife in this area is the responsibility of the Assistant Game Warden (AGW) Salt Range, who resides at Jhelum. The AGW has one game inspector posted at Sohawa to aid him. At present there is one wildlife watcher performing his duties wholly or partially with in the boundaries of the park and of the neighboring area. Major management interventions that had been wholly or partially implemented at the time that this survey was undertaken included: Development of urial enclosures; Anti poaching measures; Curtailment of livestock grazing; Development of artificial water reservoir; Development of children park; and The plantation of various species of tree, including many exotics.

Chhumbi Surla Wildlife Sanctuary

Chhumbi Surla Wildlife sanctuary, notified in 1978 has an area 55,943 hectares and is located at Lat. 32^0 50 N., Lon. 72^0 46 E, in the Chakwal District of Punjab.

A significant part of the sanctuary is underlain by dolomite and may, technically, be regarded as Karst topography. The smaller streams are essentially ephemeral; water available at several springs emerging from the dolomite aquifer (IUCN, 1997). Its core area, about 6075 hectares is state forest, located almost in the middle of the sanctuary area, mainly on the hill slopes ranging in elevation from 460 m to 1050 m above mean sea level. It is named after village Chhumbi and Surla Reserved Forests forming main bulk of the sanctuary. The area surrounding the core zone is mainly community forests, agriculture and grazing land. The Ram Halawan and Dhram Terath are two small forest patches also included in the sanctuary. Core area has good vegetation cover with open canopy suitable, not only for urial but also for a number of other vertebrate species. The community forests are poorly preserved and have been subjected to pressures including clearing of land for agriculture, livestock grazing and brushwood cutting for fuel and fodder.

The forests mainly comprise of phulai (Acacia modesta) but on higher altitudes wild olive (Olea ferruginea) and sanatha (Daytona viscosa) also exist. Other important shrub species are bhekar (Justiciar adhatoda) pataki (May tenus Rawlins) and malla Zizyphus nummularia. Important grass species include Chrysopogon serrulatus, Hetropogon contortus, Digitaria sanguinalis, Dichanthium foveolatum and Dactyloctenium scindicum.

There are at least 25 villages on the periphery or within the area. People of the villages and their livestock are mostly dependent on the Chhumbi Surla wildlife sanctuary forests for firewood and livestock grazing. There also are scattered small coal mining operations on the south-eastern side of the sanctuary.

Rakh Jalalpur Wildlife Sanctuary

Rakh Jalalpur having an estimated area of 2263 hectares is located at Lat. 32^0 41 N and Lon. 72^0 32 E. It is an intact mass of land in the Jhelum District and was notified as Wildlife Sanctuary in 1991. It has almost similar vegetation and range conditions as prevailing in Chhumbi Surla, except that Acacia modesta here is found scattered in patches on the slopes and along streambeds, often in association with Salvadora oleioides. Olea ferrugenia is totally absent.

Climate

The climate of the area is sub-humid sub-tropical continental type. The thirty-year average precipitation was 853 mm. There are two distinct rainy seasons: the summer season or the monsoon rains start by about mid July and last until the mid of September. Most of the precipitation is received during July and August. The winter rains begin in January and persist up to beginning of March. The mean monthly temperature range 5.9-38.4 c, January being the coldest and June the hottest month of the year. During winters the temperature often drops to below zero, usually in December and January.

The Punjab Urial Ovis vignei punjabiensis

The taxonomy of the genus Ovis is very confusing. Different arrangements of species and subspecies have been proposed by various authors. Roberts (1997) describes Punjab Urial as a sub-species of Ovis vignei Blyth, 1841 having three sub species:

Ovis vignei blanfordi. Type locality Bolan Pass Balochistan;

Ovis vignei vignei. Type locality Astore, Baltistan and Gilgit;

Ovis vignei punjabiensis. Type locality the Salt Range.

Roberts (1997) described the distribution of urial. In Balochistan and Waziristan, the urial inhabits the gentler slopes of the higher mountain ranges and occurs up to 2750m in association with scattered Juniperus macropoda forest in northern Balochistan.

In Kalat, the Mekran coast range and Sindh, they are found in hills up to 2500ft, with stunted Acacia senegal trees and Dwarf Palm Nannorrhops ritchieana. In the Salt Range and the lower hill ranges of southern NWFP, they are typically associated with low rounded stony hills dotted with Olea ferruginea and Acacia modesta.

In Khyber and Malakand, and Chitral they are associated with Holly Oak Quercus ilex as well as Olea ferruginea. In the extreme northern and inner Himalayan ranges associated with barren treeless regions in the lower foothills, with Hipophae rhamnoides in dry gullies.

As described by Schaller and Mirza (1974) and Roberts (1997), Punjab urial ram attains full size at an age of seven. It has long massive horns, curved outwards and backwards reaching the level of the neck in a crescentic circle. The large throat ruff, which may be white at the upper portion, hangs half way to the knees. The coat colour is dark rusty usually with white saddle mark. Often there is a blackish line on the saddle and a black side stripe.

Females have slender horns, which are as long as their ears. There is no neck ruff in females and they also don't have a saddle.

Literature Review

According to Walker (1975) there are six species in the genus Ovis. The genus has its distribution in all the higher mountain masses both in the Eastern part of Eurasia and the western part of North America (Roberts, 1997). Walker (1975) described the distribution of the six species as follow:

Ovis Canadensis: North America, north to British Columbia and, Canada and east to the Black Hills of North Dakota

Ovis dalli: North western Canada & Alaska

Ovis Ammon: Western China, Russia South to Ladakh and Nepal Ovis ammon polii, during early winter, enters Pakistan, from Chinese Turkastan across the Khunjerab in to northern Hunza, and from the Hindu Kush mountains (Roberts, 1997)

Ovis laristanica: Southern Iran

Ovis musimon: Sardinia, introduced in Europe

Ovis orientalis Iran, Afghanistan, Kashmir and Pakistan

According to Aleem (1977), Roberts (1977) three sub species, so far have been described for the species Ovis orientalis Gmelin, 1774, as follows.

Ovis orientalis vignei Blyth, 1841, Ladakh urial, found in northern areas and Chitral.

Ovis orientalis blanfordi Hume, 1877, Baluchistan urial, found in Balochistan and Sindh.

Ovis orientalis punjabiensis Lydekker, 1913, Punjab urial, occupying Salt and Kala Chitta Ranges in the Punjab province.

Roberts (1997) placed the Nearctic sheep in a separate subgenus Pachyceros and split the Asiatic 'Red Sheep' into two specie

Ovis vignei, the urial occurring from Soviet Central Asia, north-eastern Iran, Afghanistan, Pakistan, and north-west India; and Ovis orientalis, the Asiatic Mouflon, being confined to Asia Minor and western Iran. Punjab urial is a sub-species of Ovis vignei Blyth, 1841 having three sub-species:

Ovis vignei vignei Type locality Astore Gilgit.

Ovis vignei blanfordi Type locality Bolan Pass Balochistan Astore Gilgit;

Ovis vignei punjabiensis Type locality the Salt Range.

Punjab urial is the principal mammalian game specie of the scrub forest in Salt and Kala Chitta Ranges in the Punjab (Aleem, 1977; Mirza et al, 1980; Chaudhry et al, 1991). In the Salt Range and the low hill ranges in southern NWFP they are typically associated with lower rounded stony hills dotted with wild Olive and phulai (Roberts, 1997). Punjab urial are found in Pakistan distributed between the Indus and Jhelum rivers from an altitude of 250-1,500 m primarily in the Kala Chitta and Salt Ranges (Schaller and Mirza, 1974).

Feeding activity is confined to the early morning and evening (Aleem, 1977; Roberts, 1997). Their preferred food is grasses (Roberts, 1997,).The preferred species in Salt Range are Eleusine flagellifera, Digitaria bicornis and Cenchrus pennisetiformis (Schaller and Mirza, 1974). Acacia modesta is the favourite browse and Lasiurus hirsutus is also a relished grass (Mirza et al, 1980). While browsing they feed themselves on low hanging branches of Olea ferruginea, Acacia modesta and Zizyphus nummularia in Chak Jabbi area Kala Chitta Range (Aleem, 1977).Grass and grass like plants contribute up to 72% of the plants used by the big horn sheep (Constant and Kerry, 1973). Hoefs (1974), while studying the food selection in Ovis dalli, reported data from Yukan game Canada, collected throughout the year and stated that grasses and grass like vegetation contributed 59% of the forage, with forbs about 19%, shrubs 19%, shrubs and trees about 17.5% and dried bark about 3.5%.

Geist (1968) based on physical features, established different age classes for mountain sheep. Schaller and Mirza (1974) used the same procedure to classify age group classes for Punjab urial, which for rams were partly based on the horn size. The identified sex and age classes were:

Young;

Female;

Yearling male (1½ Years);

Class 1 male (2½ Years);

Class II male (3½ years);

Class III male (4½ years); and

Class IV male (fully-grown).

Body size, body proportion, horn shape, horn length, presence and size of neck ruff and colour of hair were used as criteria to distinguish the sex and age and differentiate between different age classes.

Punjab urial are said to mate in September and October (Prater, 1965). In the Salt Range the rut commences from about mid October and lasts until late November (Schaller, 1971). In southern Baluchistan and Sindh the rut may start 8-10 days earlier (Roberts, 1997). Schaller & Mirza (1974) studied the herd size and composition and found it to vary with season. While describing the agonistic behaviour Schaller and Mirza in most instances followed the terminology used by Geist (1968, 1971) and quantified 13 agonistic patterns to be present out of 20 described by him. These are clash, jump, horn pull, butt, jerk, poke, mount, low stretch, twist, kick, head down, rub and horn. Mammalian males generally tend to interact with other males of their own size and it has been reported for a number of species, including Axis deer (Axis axis) (Schaller, 1967) and mountain sheep (Geist, 1968). Cavallini (1987) studied pre-rut behaviour of Ovis orientalis musimon in Italy and found 12 behaviour patterns not to be present out of 19 considered.

Pendu-Le, et al (1995) while studying inter-individual associations and social structure in Ovis orientalis musimon observed a population of the species in Canada for one year and stated that no strong association was found among adults but some of the young females maintained a preferential bond with their mother until the age of three. Festa (1991) studied the social system of big horn sheep (Ovis canadensis) in Alberta for

10 years. He found that yearling males did not associate with their mothers. Some rams even left their native group to rut elsewhere. Shackleton (1991) studied social maturation and productivity in bighorn sheep in USA during 1988. His experiment involved two populations, a hunted one where most of the mature males were removed and the other with no hunting. One month later he found that the remaining young male courted and successfully copulated with adult oestrous females in the hunted population. However, young males in the unhunted population not only failed to copulate but also found to perform mainly immature patterns in male-male interaction. So the social maturation in mountain sheep is not simply a function of chronological age but is also related to social structure and composition of the population.

Mirza, et al (1980) surveyed the Kalabagh 'Game Reserve' in April 1976, Jhelum District in March, May & June 1976 and Kala Chitta Range in November & December 1976 to determine the status of Punjab urial and estimated a population comprising 2157 individuals. Aleem (1977) estimated 47 animals in (Chak Jabbi) a small part of the Kala Chitta Range in 1977 (Aleem, 1977). Estimates by Chaudhry (1992) give a minimum total population of 1,550 throughout its whole range and again for Punjab, Chaudhry, et al.(1988) reported a significant decline in Urial numbers over only one year; from 733 in 1986 to 528 in 1987. The population of urial, was believed to be only 60 head in the Chhumbi Surla Wildlife Sanctuary (IUCN, 1997).

Schaller & Mirza (1974) classified 1987 urial with respect to their age class and sex and described the composition of the population in Salt Range as: Males = 38.4% as compared to 37.7% females with almost a 1:1 sex ratio. Aleem (1977) described composition of the population in Kala Chitta range as: Males, 32% as compared to 57% females. The male/female ratio was about 1:2. Edge and Edge (1987) while studying ecology of wild goats and urial in Kirthar national park reported male/female ratio 0.58:1 for adult urial. A 1:1 sex ratio is common in mountain sheep populations (Buechner, 1960).

Gestation period also has not been agreed upon by various authors. Stockley (1922) reported the gestation period to be about 160-164 days. Roberts (1997) reported it to be 134 days. Walker (1975) reported that gestation period varies from 150-180 days in various species of Ovis. Prater (1965) stated that one or two young are produced at birth. Roberts (1997) reported that twin lambs are born only occasionally and this probably depends on the availability of fodder/forage during the rutting season. Schaller & Mirza (1974) reported that most ewes were accompanied by only one young suggesting that in the study area single births were the rule and not the twins. In Salt Range, Punjab, Aleem (1977) reported 0.19 young per female, which is quite lower than Schaller & Mirza (1974) i.e. 0.75. Valdez (1976) conducted fecundity studies on wild sheep (Ovis orientalis) in Iran and found that the foetal rate varied with the age of females, which on average 1.15 to 1.38 for ewes having 1-4 year age and 1.60 for ewes more than four.

Bon et al (1993) while studying the mating and lambing relationship with the age of female in Ovis gmelini in southern France found that females greater than or equal to three-year age were more productive than two-year-old females. Schaller & Mirza (1974) while studying birth and survival rates of urial in 1969 and 1973 in the Salt Range, Punjab stated that 75% of the adult females were breeding. About half of the young disappeared between the ages of 6-18 months. Festa (1988) while studying breeding and survival in big horn sheep in south western Alberta, Canada found that the viability of lambs born after 10 June was extremely low. Inadequate nutrition was found to be the cause of mortality.

When forage quality declined, the mother could not produce sufficient milk to ensure lamb survival.

Chaudhry, et al (1991) studied the mother offspring behaviour in urial under captive conditions at Lahore Zoo from March 18, 1986 to May 3, 1986 and described licking, suckling and agility of the young. Maximum suckling was observed during the first week after birth, which progressively decreased with age. Licking is essential for strengthening the bond between mother and young and enables the mother to identify and distinguish her young from the others (McFarland, 1981). Obregon et al (1992) while studying maternal expenditure during lactation in Ovis orientalis musimon at a National Park in Spain, in a captured group of seven mother infant pairs stated that males suckled for longer periods than females while unsuccessful attempts were more frequent by female infants.

Bon et al (1995) while studying the use of feeding habits by lactating female mouflon (Ovis gmelini) in Canada stated that female with lambs more than or equal to three days old fed more on rocky areas and stayed closer to safe terrain than did other mouflons which more frequently used slope tops. Festa et al (1988) while studying the anti-predatory strategy of big horn ewes in Canada stated that pregnant ewes migrated from a range of high-quality forage to one of low quality forage, apparently to avoid predation on new born lambs.

Geist (1971) found no evidence for territorial behaviour in mountain sheep. Tener (965, cf. Schaller, 1977) found no evidence for territorial behaviour in musk ox (Ovibos moschatus) nor does the literature provide positive information on takin (Budorcas taxicolor). Schaller (1977) cited that territories in the sense of an actively defended area have not been well documented among sheep and goats. Pfeffer (1967 cf. Schaller, 1977), with respect to mouflon, wrote: "It is only at the moment of reproduction that certain males choose a territory which they defend against the approach of rivals. The other males practice a hierarchical promiscuity". We noted similar behavior in urial.

Schaller (1977) wrote that it thus seems possible that in certain situations sheep exhibit territorial behavior, especially if the concept of territory is broadened from a defended area to an area of exclusive use regardless of the mechanism employed to maintain it.

Aleem (1977) and Mirza, et al (1980) reported high poaching rates for urial in Salt and Kala Chitta Ranges. Aleem (1977) has reported poaching of young ones for keeping urial as pets. Wolves, leopards and caracal cats have been observed preying on urial (Stockley, 1936 cf. Roberts, 1977). In the Salt Range, Punjab, the leopard and caracal are now very rare and human predation is probably the only significant factor, and so is the competition from domestic grazing flocks (Roberts, 1997). Kenneth et al (1978) while studying a grey wolf and stone sheep (Ovis dalli stonei) fatal predatory-prey encounter in Canada stated that the down hill chases are the most successful hunting techniques used by wolves to kill the sheep.

Materials and Method

Field Observations on Urial

Field observations were made using spotting scope 15-60x60 x and (10 x 50 and 7 x 35) binoculars. The animal was watched, while walking along ridges and ravines in the early morning and late afternoon or sitting quietly on cliffs and watching the aspects facing the observer. On spotting an urial herd, the number of individuals was counted. The age and sex were determined. During field observations it was often very difficult to determine the sex of the animal, for the yearling males, and the ewes looked, very much similar except for minor differences in their size.

Monitoring Urial Abundance

The material set out below adopted for monitoring population trends in the numbers of urial in the Salt Range, using a ground survey conducted once each year. The following need is considered in planning a survey of this kind:

1. The type of survey chosen should be feasible with limited resources, and yet provide a useful measure of the status of a population
2. Decided survey lines (fixed routes) within the Salt Range with the help of global positioning systems (GPS).
3. The nature of the terrain in Salt Range suggests that the maximum distance walked by a field survey team in a day not exceed 10 km. In view of the above, the following survey strategy is adopted:

Type of Survey

The line transect method be used, using a simple count of the number of animals detected ahead during the transect line.

(A line transect is a survey line across country, usually walked by a team of observers). This method is straightforward to use and, when used with care in a relatively uniform set of observing conditions, can identify trends in population numbers over time. The data obtained is compared with the result of the survey carried out in 1999.

Timing

One survey to be conducted annually in December.

Transect Policies

A fixed set of transects decided and used for future surveys. The results between one year and the next are to be comparable, because simple counts depend on observing conditions such as visibility, and these change from place to place.

Sampling a Mountain Range

To avoid biasing the survey towards animals in any particular part of a mountain range, mountains are sampled by means of a transect line running either from the base of the range on one side across the crest to the base of the range on the other side, or from the base to the crest of the range, then down again by another route. To minimize the sternness of the climb while maximizing the distance walked on the range, the route is climbed on the range at an oblique angle (e.g. up a ridge) and return to the lowlands in the same way.

Direction of Walking

During survey transect walked into the prevailing wind to reduce the chance of animals detecting their scent. The starting and finishing points of each transect marked by GPS coordinates.

Time of Day

A suitable time to begin transects is as early as possible in the day, and not more than 2h after daybreak, so that transect finished before dark.

Counting Procedure

All animals ahead of the observer counted from the start to the end point of each transect. Care taken not to count an individual more than once; care also taken in counting groups because, once groups are disturbed and start to move, additional animals may join the group. Binoculars used only to check identifications and to assist counts of distant groups; they not are used as a standard aid to searching.

Survey Records

After each survey, the original completed data sheets photocopied.

Summarizing Findings

Following an annual survey, the numbers of each transect totaled separately. A way of estimating densities from these data is set out below. [An alternative procedure is to divide the totals by the total length of transects walked to give a population estimate expressed as overall number per kilometer walked. A conversion factor is used to produce a rough estimate of total numbers in Salt Range. A data sheet suitable for this work has been prepared and is included at the end of this report.

Transect and Section Lengths

The overall transects length should be worked out (in kilometer) from an accurate map, then converted to meters (x 1000). Individual transect section lengths can be worked out approximately from the total transect length and the duration of individual sections by assuming that the field team walked at an even speed. Work out the total time spent on a transect (by adding the duration's of the transect sections). Calculate the length of each section by working out:

Section Length (m) = [section duration (min) x total transect length (m)] / [total time on transect (min)]

[For example, if a particular transect section took 10 minutes, the total transect length was 10 km (10,000 m), and the total time on the transect was 6hrs. 40 min (400 min), the section length was approximately (10 x 10, 000) / 400 = 250 m].

Estimating Densities

An estimate of population densities or even total populations is made for an individual transect section, a complete transect, or the full set of transects, by using the following equation:

Estimated Density (no./sq. km) = [no. Detected x 1,000,000] / [2 x distance walked x delectability Coefficient]

The delectability coefficient is a number, which summarizes how difficult a particular species, is to detect under the conditions of a census. The computer program used to estimate population densities from the helicopter data returned values for the delectability coefficient for the three large indigenous herbivores of KNP, as follows:

Sindh ibex 145

Urial 109

Chinkara 155

The lower the delectability coefficient, the more difficult it is to detect the species concerned. [For example, if a particular transect was 10, 000m long, and 50 urial were detected by the census team, an estimate of urial density on that transect is given by using the formula above. The estimated density = (60 x 1,000,000) / (2 x 10,000 x 109) = 22.9 individuals/km^2].

Estimating Total Populations

A total population can be estimated by multiplying the density estimate by the total area of the region containing the animals. [For example, suppose that in 50 km (50,000 m) of transects, 75 urial had been detected, giving a density estimate of 6.7 animals per square kilometer. The total population estimate would be 6.7 x total area = total population approximately.

Standard Errors

All survey work is subject to error, partly as a result of the sampling process itself, including the number and placement of transects, which are less than adequate for an area the size of salt range. This suggested monitoring procedure have, in addition, a certain amount of error in the delectability coefficients, which are based on helicopter surveys rather than ground surveys and in which fewer animals were detected than is ideal for modeling purposes. Nevertheless the estimates from line transect work should give park management a sufficient picture of population trends to be useful, despite their being subject to error.

Some idea of the scale of error can be gained, if desired, by calculating standard errors of the overall estimates. To do this, a density estimate can be made for each transect section and the set of density estimates produced used to calculate standard errors in the usual ways.

Making Observations

Animals ahead are counted only, from 90° to the left of the path to straight-ahead to 90° to the right. [Do not count animals you did not see until you passed them. If you want be make a note of these, do so. But make sure they are not included in the totals]. Try to give all directions ahead an equal attention.

Whenever sighted any animal stop and count the number in the group. Record the sex and the number of individuals in each group. When the animal(s) take flight and move away recount the number of individuals as they move to check that recorded the correct number. Every effort was made not to double count the animals.

While walking, noted the main habitat type around and the prevailing weather conditions. Watch the time; end transect sections if we moved from one habitat type to another (e.g. lowland stony ground to mountain slopes), or when 20 minutes has passed. At the end of a section recorded: the finishing time, the time it took to walk the last section (usually 20 minutes or less if the habitat changed), the habitat type of the last section, and the starting time of the next one. [Section length can be worked out afterwards. Keep the pace fairly slow and as even as possible. If a few minutes' break is needed, take it at the end of a transect section.

At the End

Stop counting when we reached the finishing point, even if 20 minutes has not passed. GPS used to check that we are in the correct place. If not, move to the correct pickup point. Complete (or revise) the details at the top of the data sheet, except the distance traveled.

Results

The detailed study of flora and fauna is conducted with special emphasis on the urial and its habitat in the Trans Indus River area is given. The vegetation study of Kalabagh game reserve is conducted and yet not compiled. The detailed survey of Salt Range for distribution and status is conducted while adopting the above-mentioned methodology for monitoring urial abundance. After processing the data the population estimates will be presented in the next report, along with the vegetation study of the protected areas in the salt range.

Census Strategy

The Lehri nature park was divided in to four main units to conducting urial census. Each census unit was surveyed on a different day, allocating one day to each. The Buddan Maira and rest house units, having a greater possibility of inter unit movement of urial, were surveyed on consecutive days. The entire census was repeated after an interval of twenty-two days to recheck the count.

Population age Structure

For population age structure studies, age classes were taken as proposed and defined by Schaller and Mirza (1974) and was also applied by Awan (1998).

The observed males belonged to all age groups. However the number of observed mature males belonging to class three was more as compared to any other class. The relative scarcity of the mature class IV males may be due to discriminate hunting of the class due to chance or inadequately smaller sample size.

The presence of the young one in the area suggests that the population is reproductively active. However the number of young ones in the population and the number of young ones per adult female was quite low.

The census results and the conclusion derived from them should be interpreted as base line information upon which an accurate population model may ultimately be developed for a species management.

Livestock is serious competitor of urial for food due to overlapping niches, which have generally been overgrazed, much beyond the carrying capacity levels. Thus the urial has been forced to occupy marginal habitat patches in the face of this competition.

Predation and competition with the livestock are the two important factors for the decline in the population. Predation is the most significant mortality factor. Animal predation does not, however, influence the urial population in the Salt Range conspicuously. Human predation, on the other hand, has been much more decisive.

Hunting is also a type of predation where man as a hunter also behaves as a predator. There is a long tradition of hunting in the Salt Range, and successful hunters are held in high esteem. Subsistence hunting by members of the local community is almost nil but rich landowners and Army officers were cited as the prime offenders in these violations. During interview with local community people revealed that poaching/ shooting in the area is with the league of wildlife officials. Urial is hunted for meat, for trophies and for their economic value (skin and horns).

Habitat Utilisation by Human Communities

The inhabitants of villages around the forests enjoy grazing rights over the forest. Record of settlement is annexed. Livestock rearing is the major source of their livelihood. Each house hold keeps livestock and their numbers have increased over time. No effective grazing management system is being used and all the livestock is grazed openly in most of the area. In the Lehri

Nature Park area little stall feeding is practised. Each household unleashes their cattle once they have milked them. The cattle roam about in the near by village forest.

The practice of gathering fuel wood by villagers has continued for hundred of years. Herdsmen also cut the smaller branches of trees, especially Acacia modesta and Zizyphus nummularia for forage for livestock during winter months. Allowing this practice to continue will alter the natural balance of the affected ecosystem and, therefore will impede the protection and preservation of the fauna and flora in its natural state. Grazing and gathering of fuel wood has been such a long traditional use that to eliminate or even curtail its use will be difficult, considering the associated social and political constraints.

5

Urial Habitat in Rakh Topi Game Reserve and Boraka Wildlife Sanctuary

INTRODUCTION

Pakistan can be divided physiographically into four regions: the Great Highlands, the Balochistan Plateau, the Indus Plains and the Desert Areas. The Himalayan and the trans-Himalayan mountain ranges rising to an average elevation of more than 6,000 m include some of the world's highest peaks, such as K2 (8,611 m) and Nanga Parbat (8,126 m).

The country lies at the western end of the South Asian subcontinent, and its flora and fauna are composed of a blend of Palaearctic and Indomalayan elements, with some groups also containing forms from the Ethiopian region. Indomalayan forms are found in the east of the country, in the Indus Basin, and Palaearctic forms in the mountains of the north and west. The blending of elements from these different origins has ensured a diverse and unique mix of flora and fauna (Kohat Gazetteer 1990).

Pakistan is having a variety of habitats from seashore and deserts to high mountainous areas in the north. Within its boundaries it includes four phytogeographical regions namely Saharo-Sindian, Irono-Turanian, Sino-Japanese and Indian. They help to explain the richness and diversity of its flora. Most of the area of Pakistan is characterised by the Saharo-Sindian region, which includes Sindh, southern Punjab, southern Balochistan and plains of NWFP (Nasir and Rubina 1995).

Rakh Topi Game Reserve and Boraka Wildlife sanctuary, these two protected areas supports a thin population of urial. Rakh Topi Game Reserve and a part of Boraka Wildlife sanctuary lies in district Kohat. It is comprised upon Tehsil Lachi that is sandwiched between Tehsil Kohat in north and district Karak and Mianwali on south. Towards the east it is separated from district Attock by the mighty Indus, while district Hungu is located in the west.

Lachi town is situated in a broad valley between Swanai Sar and Gungalot Hills range running from west to east. This wide valley forms the Lachi plain on western end (Surgul and Darmalak), while eastern portion of the valley is known as Malgin Plain (Soodal). Further more towards the east another valley known, as Shakardara is located. Beyond this point towards the east, area is rugged, dry and barren plateau. This portion towards the Indus gives place to low ranges of broken hills with varying heights. The height of town Lachi, the Tehsil Headquarter is 519 m above sea level (Kohat Gazetteer 1990).

PHYSICAL FEATURES/TOPOGRAPHY

Mountain and hills dominate topography of the area. The rocks forming the hills belong to the tertiary formations, consisting of sandstone and limestone, tilted throughout the area at a very high angle. The rocks are not pure and mixed with shale, clay or sand. The sedimentary configuration of the rock strata is highly susceptible to water erosion especially on vegetation free hilly slopes. Erosion is further accelerated due to unchecked over grazing of the area on the lands, shamilats as well as on the state owned forest areas. Thus the grazing incidence being higher than the area can support is deteriorating the situation day after day. The area comprises of highly undulating surfaces and cultivation is possible only on flat or moderately sloping lands.

The height of these ranges varies from 633-1000 m above the sea level. The hills gradually rise in the extreme northeastern part of the study area towards the Indus River. The intervening open valleys between the hills are seldom more than 8 km in width.

The northern and southern halves of the study area differ altogether in geological characteristics. Generally speaking, bulk of the geological strata consists of sandstone and earthy conglomerate. It is divided into northern part that is predominantly rich in limestone that crops up out of the alluvial valleys. These rocks of which they are composed are generally very hard. In the southern part, the higher ranges are made up of sandstone and hard rocks of gypsum series that form the backbone of the higher ranges. The special features in the geology of the area are the enormous beds of rock salt.

EDAPHOLOGICAL CHARACTERISTICS

Texturally the soils of the study area vary from sandy loam to clay loam. The characteristic feature of the area is red soil, the red marl, which imparts red colour to the soil is dominant. The electrical conductivity ranges from 0.01 to 3.2 dSm^{-1} with average of 0.35 dSm^{-1}, while pH varies from 7.1 to 8.8 with mean value of 7.7. The contents of $CaCO^3$ and organic matter range from 2.0 to 24.0%, 0.03 to 3.07 % with averages of 9.73 and 0.82 respectively. The phosphorus content ranges from 1 to 56 ppm with mean of 8 ppm while the potassium levels range from 70 to 556 ppm with average of 174 ppm (ARI Tarnab).

Climate

The climate of the project area is arid/semiarid, hot and dry. It is terribly hot from May to September, where June and July are the hottest months. The mean maximum and minimum temperatures range from as low as 2.1°C in December/January to as high as 41°C during June/July. Average maximum and minimum temperature is 32.43°C and 14.11 °C respectively whereas highest and lowest temperature ranges for each month.

The winter is cold and severe. In winter a strong chilling west wind known as "Hungu Breeze" often blows down from the Meranzai valley towards Kohat. The rainfall is received through out the year, but its occurrence is unevenly distributed during most part of the year, and highly unpredictable. However, monsoon rain is received from July to August, which are the rainiest months.

The maximum relative humidity has been recorded in the month of July during summer season and in December during the winter season for the year 2000. In Lachi less than 45 % of the total cultivable area is irrigated. Most of the area is non-cultivable due to hill topography and depends entirely upon the rainfall. The hills are of typically arid and semi arid nature.

River and Streams

The River Indus forms the eastern boundary of the district as well as study area, which separates it from the province of Punjab. Study area has two main streams, Teri Toi, which flows from west to east in the southern half of Study area, which finally joins the Indus close to Shakardara. Logari Algad flows from south to north and drains into Teri Toi. Lachi Toi drains Union Council Lachi and joins Kohat Toi near Shadi Khail.

Besides these perennial and non-perennial streams there are several small hill torrents and tributaries, which join Teri Toi and Kohat Toi at various points while passing from study area. Among these streams, important ones are Summari Toi and Sari Algad (Soorgul), Parnazai Toi, and Ghorzandi Algad (Darmalak), Khakh Algad (Soodal) Pakka Algad and Nareri Algad (Shakardara). In general the direction of the water shed in the study area is northwest to southeast. A general tendency is noticed on the part, both of the mountain ranges and of the rivers of study area to converge towards the east.

Socio-economic Environment

The study area lies in the district of Kohat, which is counted among the more developed districts of the NWFP province. Most of the project area, however, remains largely untouched from the development work point of view and traditional sociopolitical and economic system prevails.

Livestock

The livestock of study area includes cattle, buffalo, goat, sheep, camel and poultry. The local cattle breed in study area is Lohani Dani, which is crossed with Frezein breed to increase the milk and meat production. Similarly the local goat breed variety has also been improved. Most of the people in the study area own goats, sheep and cow and they graze openly.

Study Area Management

The NWFP Forest and Wildlife Departments manage the area. Major management interventions that have wholly or partially been implemented at the time that this survey was undertaken included:

Anti poaching measures;

Curtailment of livestock grazing;

Development of artificial water points; and

The planting of various species of tree, including exotics.

HABITAT TYPES

The dominant habitat type in the area is dry temperate semi-evergreen scrub forest (Roberts, 1991).

Study area is almost half of the District Kohat's total area. The scrub forest covering the hills is dry sub-tropical semi-evergreen in nature. Because of the hill topography study area presents a variety of habitats, which inhabits unique flora. These habitat types in the area correspond with the various landform features. The area consists of vast flat plains traversed by several dry streams originating from low hills. A chain of low hills extends from the West Bank of the River Indus in to Balochistan. In all, there are seven main habitat types, each supporting different communities, which are discussed below.

Flat Plains

The study area includes extensive flat alluvial plains and some with small undulations covered by relatively similar vegetation, mostly small trees and dwarf shrubs. Tall, clump-forming grasses are common.

Species diversity is low and large areas are covered by the same species of shrubs and trees. Common trees include Acacia modesta, Zizyphus nummularia, Capparis aphylla, etc. Some cultivated trees, well established around villages and towns, include Eucalyptus globulus, Dalbergia sissoo, Prosopis juliflora, Broussonetia papyrifera etc. Common shrubs are Adhatoda vasica, Grewia tenax, Gymnosporia royleana, Calotropis procera etc. Common grasses found in this habitat include Cymbopogon jwarancusa, Heteropogon contortus, Aristida adscensionis, Cenchrus ciliaris and Cynoglossum lanceolatum etc. In this habitat five stands were selected for vegetation sampling.

Variations in soil texture and salinity have given birth to a number of microhabitats, which house slightly different plants. The microhabitats include sandy plains, rocky or gravelly sites, and saline patches. It is difficult to give clear-cut definition for each of the microhabitats in term of plant species. Nevertheless, they can be distinguished on the basis of overall total vegetation cover, dominant species, and diversity of species.

Saline Plains

Saline areas are characterized by a high percentage of salts in the soil, which appears as a white crust on the ground. The degree of salinity varies; the more intermediate habitat types, such as sandy saline patches and saline waterlogged soils near streams also support vegetation with different combination of plant species. Mostly this includes grasses with varying degree of presence.

Sandy Habitat

The soil in this habitat type contains a high percentage of sand and has a low water retention capacity. The main species found in this habitat is Saccharum munja. This is accompanied by several species of annuals like Boerhaavia procumbens and creeping grasses, such as Cynodon dactylon, Chloris sp and Asparagus adscendens. Certain xerophytes found in this habitat were Opuntia dillenii and Aloe vera. Three stands were laid in this habitat.

Rocky or Gravelly Habitat

There are often no trees in these habitats; wherever they occur, they are sparse. Vegetation consists largely of low growing shrubs. Common species include Otostegia limbata, Gymnosporia royleana, Saccharum munja and Periploca aphylla. Considerably large portion of study area is composed of this habitat.

Piedmont Grasslands

Piedmont grasslands are those that occur on the gently undulating plains at the foot of higher ridges (especially Chattru Maila/Oot Maila) in Rakh Topi Game Reserve. The grasses that include Cymbopogon jwarancusa, Heteropogon contortus and Cenchrus ciliaris dominate vegetation of such habitat. Saccharum munja is found frequently between these low land ridges in depressions.

The piedmont grasslands provide habitat to a spectrum of wild mammals, notably the Asiatic jackal, fox, wild hare, porcupine, hedgehog and several other small mammals. The reptile species also occupy this habitat along with many birds including larks, nightjars, sparrows, and other passerine birds.

Seasonal Streambeds and Banks

Several scattered streams are found in the area. Examples are the Teri Toi, Lachi Toi, Sumari Toi, Kohat Toi and Logari Algad. Most remain dry for the greater part of the year and bring water and fresh soil from the hills to the plains in the rainy season, causing occasional flash floods. Some of the streams are fed by underground springs. Some have brackish water with a varying degree of salinity. Wherever some soil and moisture is available along the streambeds and banks, a variety of plants including Cyprus rotundus, Adiantum capillus-veneris, Adiantum incisum, Aristida adscensionis, Chenopodium album, Convolvulus arvensis and Conyza bonariensis are supported. Dry streambeds support different vegetation species from the adjoining plain areas. This includes perennial grasses and among that Saccharum munja is abundantly present.

Seasonal or ephemeral streams are a source of drinking water for many of the animal species of the area. They also provide essential moisture to the plants growing on their outwash plains. Most of the wildlife species found in the piedmont grasslands and foothills are also found in discontinuous patches of seasonal stream beds and bank habitat type. Two stands were selected here to study the floral diversity.

Croplands and Villages

Any settled village is often a site centered on land irrigated year round, with cultivated vegetable and cereal crops, shrubs and trees. This habitat type represents a variety of irrigated and dry land agricultural fields. Typically the fields are located at lower elevations in association with the grasslands. The fields may be distinguished by regular margins and borders, and the overall geometry of the landscape. Irrigated crops include guava, and vegetables, while the dry land crops consist of maize, groundnut and oil seed crops etc. Large trees often surround irrigated fields, but dry land fields are typically fenced with thorn trees and shrubs.

VEGETATION

The Study area (District Kohat) is located at the borders joining Saharo-Sindian region at the east and Irano-Turanian region in the west forming an ecotone, where its vegetation is representative of both the phytogeographical regions (Nasir and Rubina 1995). Climatically this area is harsh like Afghanistan, characterized by the hot dry summers and mild winters. The forests mainly comprise of Acacia modesta but wild olive, Olea ferruginea, gurgura, Monotheca buxifolia and sanatha, Dodonaea viscosa also exist. Forest vegetation of the area is dry subtropical semi-evergreen scrub forest type with an open canopy of Olea ferruginea, Monotheca buxifolia and Acacia modesta trees. Other important shrub species are Adhatoda vasica, Otostegia limbata and Rhazya stricta. Important grass species include Cymbopogon jwarancusa,

Heteropogon contortus, Aristida adscensionis, Chloris sp, Digitaria bicornis and Cynodon dactylon. Check list of flora of study area for summer season having 157 species belonging to 53 families.

Methods

As it was not possible to survey the entire study area in the time provided for the study, a sample of locations was selected for survey. A detailed, hierarchical process was designed to select these locations in a way that retained scientific rigor, reduced selection bias and maximized spatial and environmental coverage.

In the first stage of this process the study area was divided into five sectors/blocks for survey purposes. The boundaries of each sector were determined from extrapolating areas from 1:50,000 scale topographic maps. Within each of which at least six points were randomly located. Of these, the first four were selected for the establishment of quadrats and the remainders were retained as supplementary locations that could be used if any of the four points were considered insufficient. In addition to these, other sample points were selected to incorporate rare or linear habitat types that were too small to classify into landscape units. The selected sites within the study area were visited extensively several times in the scorching heat of summer. Unidentified plants in the field were collected and brought to the herbarium of Quaid-i-Azam University, Islamabad. Flora of Pakistan by Nasir and Ali (1972) was consulted and followed for the identification and nomenclature of the plant species.

Selection of Sample Stands

Prior to sampling, the selection of sample stands is important. For this study, 26 stands were selected for sampling.

Vegetation study site were selected on the basis of differences in ecological parameter such as soil type and texture, slope, aspect etc. in order to make the samples representative of the area.

Suitable size of quadrat is important for vegetation sampling. It is selected according to the size and spacing of the plants. Based on the visual observation of vegetation structure, the quadrat size of 1×1 m^2 was selected for herbs and grasses while quadrat size of 5×5 m^2 was used for forest plants and trees/shrubs. Quadrats were laid out at each study site at random, 50 m (90 "Karam") apart from each other. A total of 380 quadrats in 19 stands were taken for herbs and grasses, whereas 7 stands comprising of 70 quadrats were laid for the study of trees and shrubs.

Vegetation Analysis

Vegetation analyses were conducted in two different ways:

Importance Value

Twinspan Analysis

Importance value

Data for density, frequency and cover were recorded and importance value of each species calculated.

Importance Value = Relative Density + Relative Frequency + Relative Cover.

Plant communities were established on the basis of dominant species having a larger importance value (IV) in the given ecological community. A detail of parameters used for the vegetation structure analysis is given below.

Density : Density relates to the number of plants rooted within each quadrat. The sum of the individuals per species was calculated for the total area sampled by the small quadrats. The average density per quadrat of each species was converted to quadrat size of 1m^2.

Density = Number of individuals of species in all quadrats

Total area sampled

Relative density

= Total number of individuals of a species x 100

Total number of all individuals of all species

Frequency: It relates to the percentage of total quadrats that contain at least one rooted individual of a species.

Relative Frequency = Frequency of a species x 100 Sum frequency of all species

Results

Community Analysis : Vegetation of study area was analysed by two different methods: Importance value and Twinspan analysis.

Importance Value (I.V.)

Stand 1. Chloris sp Community

This stand was studied at the foothills and mid slopes, which gradually stretched into hills. Due to grazing and stone cutting habitat biodiversity was considerably disturbed. The dominant species was Chloris sp with an importance value of 52.64. The other co-dominants were Cymbopogon jwarancusa with 42.91 I. V. The least occurring species were Peganum harmala and Sida alba both with 2.89 I.V. and Achyranthes aspera with 2.66 I.V. A total of 26 species were recognized in this stand.

Stand 2. Cymbopogon Jwarancusa Community

This area was comprised of large plains thickly covered with Cymbopogon jwarancusa like a carpet on the floor. Shrubs along with Acacia modesta were often present scattered among the grasses. Quadrat analysis shows Cymbopogon jwarancusa as the dominant species with 90.11 I.V. and Chloris sp with 43.03 I.V. The other co-dominant species found were Cynodon dactylon with an I.V. of 32.44 and Acacia modesta with 17.01 I.V. The least occurring species were Salvia aegyptiaca and Randia tetrasperma both with 2.18 I.V. A total of 21 species were recorded in this stand.

Stand 3. Cymbopogon Jwarancusa Community

This site was selected on road towards Shakardara before check post. Both the slopes were studied during quadrat analysis.

The area is consisting of low height hills, which rise from plain of Shewaki and extends towards the Indus. Teri Toi drains the area.

The highest importance value found in this stand is that of Cymbopogon jwarancusa i.e. 120.41 and Rhazya stricta, 61.13. Saccharum munja was found co-dominating with an importance value of 57.03. A total of 17 species were recorded in this stand.

Stand 4. Cymbopogon Jwarancusa Community

This portion lies at the extreme east corner of Union Council (UC) Shakardara, where it stretches to the Indus. This part of UC has rugged, jagged, irregular and weather-beaten topography. During the study climate was severely hot and dry.

From Shaheed Chowk to Sharki, this entire area is composed of dry, dull mountains. Erosion is common scenery, which is facilitated due to the presence of sandstone in the topography of the area. North facing slope is richly covered with Cymbopogon jwarancusa and scattered Acacia modesta.

The dominant species found were Cymbopogon jwarancusa with 134.65 I.V. and Heteropogon contortus with 43.89 I.V. respectively. A total of 14 species were recorded in this stand.

Stand 5. Cynodon Dactylon Community

Rakh Topi game reserve is protected area for fauna and flora. At the west bank of Indus River, it is a wide flat plain, at the west of which hills of gentle slopes are located thickly covered with different grass species. This area inhabits a large variety of flora with different microhabitats. Close to the human settlements tree cover is reduced due to over exploitation of wood for fuel consumption. There were herds of goats and sheep almost in every house, which are taken to the fields in the morning and brought back in the evening. Similarly donkeys and camels are also common which are used for carriage. Gentle sloping hills in the west are important fodder ground for the cattle. The grass grown there is not only used daily but also cut and dried for winters.

Cynodon dactylon (99.91 I.V.) along with Acacia modesta (38.29 I.V.) and Boerhaavia procumbens (29.15 I.V.) were found dominant. The co-dominants were Cleome brachycarpa with 26.71 I.V. and Tribulus terrestris with 22.53I.V. The least occurring species was Digitaria bicornis with 3.88 I.V. A total of 16 species were found in this stand.

Stand 6. Acacia Modesta Community

In this stand Rakh topi area between the villages Janak and Sharki were studied for tree and shrub cover at the hillside. Once the hillside of this area was thickly clothed with Acacia modesta and Capparis aphylla (Chaghtai and Yusaf 1976). During the present study Acacia modesta was found in less numbers while Capparis aphylla was present rarely and in small bushy habit.

The dominant species was Acacia modest with an importance value of 67.98. It was the dominant tree species occurring in this stand. The other co-dominants were Heteropogon contortus and Rhazya stricta with 55.89 and 48.44 I.V. respectively. A total of only 8 species were recorded in this stand.

Stand 7. Cymbopogon Jwarancusa Community

This stand was selected at the west of Rakh Topi in Sera Maila to assess the biodiversity of grasses. The most evident grass on these gentle sloping hills is Cymbopogon jwarancusa. Saccharum munja was as usual present in the Nullah beds. Acacia modesta and Capparis aphylla were inhabited in the ravines. Grazing is not common in the area, thus the grass is protected which would then be used for fodder in winters. However, cutting at a small scale was observed to feed the cattle. Cymbopogon jwarancusa with 162.69 I.V. and Rhazya stricta with 59.81 I.V. were found dominant. Chloris sp with 33.14 I.V. was found as co-dominant species. Only seven species were recorded in total in this stand.

Stand 8. Cynoglossum Lanceolatum Community

This site is located in south of the Rakh Topi and close to the River Indus. This plain area gradually rises to hills and finally becomes mountain more towards west. These ranges run along the river Indus and stretch to the villages Rukwan and Mullawalli. Hills are gentle and covered with grasses. Cutting of wood is common and grazing is very severe.

Cynoglossum lanceolatum with 105.62 I.V. and Chloris sp with 51.79 I.V. were dominantly present. The co-dominating species was Boerhaavia procumbens with 39.65 I.V. The least occurring species was Grewia tenax with an importance value of 2.95. A total of 14 species were recognized in this stand.

Stand 9. Acacia Modesta Community

In the south of Rakh Topi, a wide undulating plain with trees like Acacia modesta and Zizyphus nummularia is located. This piece of land is declared protected area for the fauna and flora.

Due to protection and carefulness this area gives different picture as compared to the rest of the area. Forest department has also done plantation of Acacia modesta which will be further helpful for the preservation of habitat in Rakh Sobar.

Acacia modesta was found as dominant tree species with 62.21 I.V. Heteropogon contortus, Cymbopogon jwarancusa and Saccharum munja were co-dominating with Acacia modesta. A total of 7 species were recorded in this stand.

Stand 10. Chloris sp Community

Steep mountains and rugged topography characterize the area. Rocks are composed of sandstone. The sampling site is located at the River Indus. The vegetation cover has shrunk due to human intervention to the remote and difficult areas and mostly close to the peaks. Different deleterious factors imposed by human population are very intense due to which the floral biodiversity is in highly pathetic condition.

Among the plants Capparis aphylla, Acacia modesta and Gymnosporia royleana are thinly present in the area though during study they could not be sampled. Ravines are covered with thick Saccharum munja. Cymbopogon jwarancusa is present on mountain slopes.

Chloris sp was found dominating the ground layer with an importance value of 68.23. The co-dominants were Saccharum munja with 51.04 I.V. and Rhazya stricta with 42.78 I.V. A total of 11 species were recorded in this stand.

Stand 11. Cymbopogon Jwarancusa Community

The present site is located between the Mukhad Sharif and Rukwan at the right bank of River Indus. Cymbopogon jwarancusa and Saccharum munja community is dominant at the lower and mid ridges while upper slopes of the mountains are devoid of any prominent vegetation except Rhazya stricta. Human intervention and dependency upon the natural resources is very severe, which is more evident near to the human settlements and villages.

Cymbopogon jwarancusa and Rhazya stricta were found most dominating with an importance value of 176.35 and 85.07 respectively. Saccharum munja was found as co-dominant with 55.16 I.V. A total of only five species were recorded in different quadrats of this stand

Stand 12. Cymbopogon Jwarancusa Community

This stand was selected in the extreme north west of study area in Union Council Soodal. Geology is porous. Gravels and boulders are frequent and erosion is heavy. Although grazing and grass/wood cutting is severe but due to the richness of palatable species and grass cover, this area is rich from fodder point of view. Cymbopogon jwarancusa was found most dominating with 166.32 I.V. The other co-dominants were Heteropogon contortus with 37.21 I.V and Monotheca buxifolia with 21.49 I.V. The least occurring species in the stand were Acacia modesta, Boerhaavia procumbens and Cynoglossum lanceolatum all with the similar I.V. i.e. 3.28. A total of 13 species were recorded in this stand.

Stand 13. Chloris sp Community

This site was selected to assess the pattern of variation and diversity of vegetation in the adjacent land of Darwazai Banda. Due to presence of water reservoir, moisture in adjoining land is high. Soil is sandy loam with open porous constitution and area is being used for many cultivated crops. Irrigation is also practiced with surplus water from the dam.

Chloris sp (73.75 I.V.) was found to be most dominating species with Cymbopogon jwarancusa having 43.12 I.V. The species co-dominating in the stand were Saccharum munja (22.64 I.V.), Zizyphus nummularia (22.28 I.V.) and Heteropogon contortus (22.14 I.V.). Lactuca orientali was the least occurring species. A total of 17 species were recorded in this stand.

Stand 14. Cymbopogon Jwarancusa Community

Red marl is a common feature of soil mostly towards the west and northwest of Tehsil. It was observed as red patches on gray and green colored soil. It sustains a different community than the rest of vegetation. Type of flora on such soil is same but their frequency and abundance are different. This type of soil is very heavily eroded due to sandy silty texture and steepness of slope that ultimately results in less vegetation cover. Soil constitution is sometimes porous or spongy or compact. Only grasses were found growing in this area.

Cymbopogon jwarancusa was found as most dominating plant species showing the highest importance value (164.13) out of a total of 4 species recognized in the stand. Heteropogon contortus was also dominant with 70.93 I.V.

Stand 15. Heteropogon Contortus Community

This stand was selected in the suburbs of Lachi town in Manduri and sampling was carried out from the foothills to the top of the hills. Grasses were highly abundant mostly above the wide ridge of the hilltops whereas shrub cover was more dominant at the bottom of the mountain. Stone mining for building is common. People often use plants (Adhatoda vasica) for fuel purposes.

The most dominating plant species that was observed in this stand was Heteropogon contortus with 92.91 I.V. The co-dominating species were Cymbopogon jwarancusa with 40.29 I.V. and Chloris sp with 30.46 I.V. The least occurring species were Aerva javanica (6.28 I.V.) and Zizyphus nummularia (3.15 I.V.). A total of 14 species were recorded in this stand.

Stand 16. Poa Annua Community

This site was once decided to be used for the construction of dam due to suitability of landscape. A reasonably thick vegetation cover of Monotheca buxifolia, Dodonea viscosa and scattered Olea ferruginea is present on the mountains due to mesic conditions and comparatively more rain. This area is also a good grazing ground. Fuel wood collection was a common sight and daily donkey carts full of Dodonea viscosa are sold in the market. Poa annua and Saccharum munja were found dominant in the quadrats showing high importance values. Cymbopogon jwarancusa and Dodonea viscosa were found as co-dominant species in this area. A total of 21 species were recorded.

Stand 17. Dodonea Viscosa Community

Dodonea viscosa was dominant at foothills while Monotheca buxifolia and Acacia modesta were dominant at mid slopes, whereas the hilltops were covered with grasses. Overall vegetation cover is thick. Prosopis juliflora was commonly growing along the road and as hedge tree. Adhatoda vasica was also observed along the roads sides. South facing slopes, at the left of Jabbi and Barh villages were entirely devoid of any vegetation because of human settlement in the vicinity except Cymbopogon jwarancusa, and Heteropogon contortus.

The dominant species observed in the quadrats were Dodonea viscosa with 74.39 I.V. and Cymbopogon jwarancusa with 56.09 I.V. The co-dominant species of the stand were Monotheca buxifolia (21.48 I.V.), Heteropogon contortus (20.52 I.V.) and Acacia modesta (20.45 I.V.). The least occurring species was Teucrium stocksianum with 3 I.V. A total of 17 species were recorded in this stand.

Stand 18. Dodonea Viscosa Community

The present stand was taken in Boraka area, which is a protected reserve forest owned by the government and set aside as an undisturbed breeding ground for the protection of wild life. Quadrats were taken on north facing slopes on the mountains running from east to west. Due to protection and care emphasized by forest department, floral cover is fare, though cutting and rooting up of plants (Dodonea viscosa) is frequently practiced for fuel requirements.

The Boraka range of mountains takes it start from Boraka village (northeast), Sumari (east) and Kamardand (southeast) and stretches in length up to the district Hangu in the southwest direction. The whole area is protected to preserve the wild fauna and flora of the area. Part of the range close to the villages is community land where human intervention is controlled and checked by the villagers. Towards the east vegetation is mostly bushy and thick. As we move on the range to the southwest, habitat changes to trees and towards Mir Khaili number of Olea ferrugenia trees increases.

The dominant species were Dodonea viscosa with 128.58 I.V. and Monotheca buxifolia with 56.53 I.V. The co-dominant species was Cymbopogon jwarancusa with 37.4 I.V. A total of only 9 species were recorded in this stand.

Stand 19. Dodonea Viscosa Community

This site is a narrow belt of flat plains between the mountains running from east to westward direction located between the Sumari Bala and Sumari Pai· n. As described earlier this area joins with the Boraka Reserve Forest in the west, and comprises of rich vegetation.

Grazing, grass cutting and fuel woodcutting is common practice. Vegetation is rich on both the slopes due to protection measures, whereas, northern aspect is more generous to support healthy and varied plant communities.

Camel, sheep, goat, cow and donkey are the most common domestic animals reared on this land. All of them freely graze in the area.

Dodonea viscosa with 94.49 I.V. and Saccharum munja with 41.04 I.V. were found dominating the area. The species co-dominating with these were Heteropogon contortus with 37.5 I.V. and Cymbopogon jwarancusa with 32.39 I.V. A total of 15 species were recorded in this stand.

Stand 20. Cymbopogon Jwarancusa Community

In this locality there were mountains which were devoid of vegetation above mid slope whereas below mid hill and ravines were thickly covered with Cymbopogon jwarancusa and Saccharum munja.

Cymbopogon jwarancusa with 168.02 I.V. was found dominating the area. Saccharum munja with 28.32 I.V. was the co-dominant species. A total of 12 species were recorded in this stand.

Stand 21. Teucrium Stocksianum Community

This stand was selected to compare the immediate difference between the vegetation of protected and non-protected areas. The fenced area is mostly naturally restored with native vegetation of the area which includes Dodonea viscosa, dominating with Cymbopogon jwarancusa. Monotheca buxifolia and Acacia modesta were more concentrated towards the upper ridges. Out side the fencing, vegetation is severely damaged because of unchecked grazing and wood cutting.

Teucrium stocksianum with 67.88 I.V. and Dodonea viscosa with 64.31 I.V. were dominantly present. The other co-dominants were Heteropogon contortus with 51.87 I.V. and Cymbopogon jwarancusa with 43.52 I.V. the least occurring species in this stand was Otostegia limbata showing the lowest importance value. A total of 11 species were recorded in this stand.

Stand 22. Dodonea Viscosa Community

This stand was studied on the hills, which are located between Lachi and Sumari. South facings slopes are badly damaged while northern slope is covered with grasses and shrubs. In this stand Dodonea and Prosopis was present in the foothills while Monotheca was present on the ridges having moderate to small cover.

Dodonea viscosa was found to be the most dominating plant species in the ground layer with an importance value of 85.39. The dominant tree species were Prosopis juliflora and Monotheca buxifolia with 39.66 and 39.08 I.V. respectively. The co-dominant species was Cymbopogon jwarancusa with 43.9 I.V. The least occurring species of this stand was Heteropogon contortus with 28.1 I.V. A total of 7 species were recorded in this stand.

Stand 23. Monotheca Buxifolia Community

Towards the west of the study area, vegetation cover is better than the east and south. The present stand was studied in the Union Council Soodal at the back of village Kamardand. Although Monotheca buxifolia is present throughout the study area, in this area it was in abundance.

The most dominant tree species in this stand was Monotheca buxifolia with an importance value of 72.19. Cymbopogon jwarancusa and Adhatoda vasica were the co-dominant species with 68.15 and 58.3 I.V. respectively. A total of only 7 species were found in this stand.

Stand 24. Dodonea Viscosa Community

This site was studied on the gentle sloping hills between Lachi and Darmalak. The hills become steep towards the top but the hilltop was flat.

Dodonea viscosa dominated all the other species recorded in the area along with Cymbopogon jwarancusa with 109.4 and 98.95 I.V. respectively. Among the trees Monotheca buxifolia (65.83 I.V.) was dominant. A total of only 4 species were recorded in this stand.

Stand 25. Monotheca Buxifolia Community

This site was selected between Darmalak and Ghurzandi. The area was composed of high ridges more covered with tree species. Monotheca buxifolia was the dominant species with 58.55 I.V. The other co-dominants were Prosopis juliflora (55.94 I.V.) and Dodonea viscosa (51.33 I.V.). A total of 7 species were recorded in this stand.

Stand 26. Dodonea Viscosa Community

This stand was selected at the end of Sumari Bala on right of the road on south facing slope. The vegetation at the foothills on gentle slopes was thick which decreased gradually towards the top of hill.

The most dominating species of this stand was Dodonea viscosa with an importance value of 96.69 and Heteropogon contortus was found as co-dominant with 75.77 I.V. Monotheca Buxifolia was also present. A total of only 6 species were recorded in this sampling site.

Twinspan Analysis

The vegetation analysis through twinspan separates the quadrats due to certain species that do not occur in common or it results in grouping up those quadrats that show common features.

The data collected during phytosociological study of study area consisted of 26 stands through out length and breadth of the Tehsil. In 19 stands 20 quadrats each of 1 m^2 were placed where the vegetation was grassy or herbaceous. Those places where vegetation comprised of trees and bushes, 10 quadrats each of 5 × 5 m^2 in 7 stands were placed which together made a total of 450 quadrats in 26 stands. A total of 75 plant species were identified in the stands. The species were arranged alphabetically and assigned numerical codes. The cover values of each species in each quadrat were then expressed in terms of Domin's Values.

The results obtained as shown in Fig. 31 clearly indicate that at the first level Twinspan divided the whole vegetation of the area studied into two major classes.

1: Major class 1 with 375 quadrats out of a total of 450 quadrats.

2: Major class 2 with 75 quadrats out of a total of 450 quadrats.

Analysis of the major class 1 shows that this community consists of the maximum number of quadrats which indicate very less variation in the vegetation of the study area areas and expresses an over all uniformity. The dominants of the class are Cymbopogon jwarancusa Boerhaavia procumbens

Chloris sp

The community therefore can be named after the dominants as.

Cymbopogon-Boerhaavia-Chloris Community

1. The Cymbopogan-Boerhaavia-Chloris community This major community can be further classified into sub communities 1-1 and 1-2. Consists of 88 quadrats out of a total of 375 quadrats of community 1. The sub communities of community one can be named exactly like its parent community, as the dominants are the same i.e. Cymbopogon-Boerhaavia-Chloris community. The other sub community of the major community i.e. 1-2, was differentiated only due to high cover value of two species, (236 and 256) i.e. Heteropogon contortus and Saccharum munja. The sub community was therefore, named after the dominants as Cymbopogon-Heteropogon-Saccharum community.

This consists of 287 quadrats in total.

The sub community 1-1 i.e. Cymbopogon-Boerhaavia-Chloris was further classified into 1-1a and 1-1b.

The class 1-1a consists of only one quadrat, which is separated from the rest due to a high cover percentage of a species Eragrostis poaeoides, however, rest of 87 quadrats out of a total 88 are a true representative of the parent community Cymbopogon-Boerhaavia-Chloris, i.e. 1-1b.

The sub community 1-2 i.e. Cymbopogon-Heteropogon-Saccharum community was further classified into 1-2a and 1-2b.

1-2a. The sub community 1-2a consists of the commonly occurring Cymbopogon jwarancusa in addition to Heteropogon contortus. Hence the absence of Saccharum munja has made the difference thus separating 237 quadrats out of a total of 287. In these quadrats Saccharum munja is either completely absent or is rare if it occurs. The community thus could be called as Cymbopogon-Heteropogon community.

1-2b. This community is representbed by 50, out of 287 quadrats and consists of a high cover of Saccharum and relatively less cover of Cymbopogon. In addition (226) Dodonea viscosa has acquired a dominant status in the community which can, therefore, be called as,

Cymbopogon-Saccharum-Dodonea Community

The major community 2 separated by twinspan consisted of only 75 quadrats. These were the few exceptional quadrats showing considerably large difference from rest of the area. The species responsible for this difference were Acacia, Dodonea and Monotheca that did not occur in the rest of the quadrats. These 75 quadrats consisted of relatively high cover of these species.

Acacia-Monotheca-Dodonea Community

Major community 2 is differentiated from the first due to the presence of Acacia modesta and Monotheca buxifolia species. It was further classified into:

Acacia-Dodonea community and

Monotheca-Prosopis-Dodonea community.

Acacia-Dodonea Community

This group consists of 40 quadrats out of total 75. The presence of Acacia modesta differentiates it from the other group. The sub community showed the same dominance, therefore, named after their parental community. It is further subdivided into 2 sub communities. 2.1a Acacia-Dodonea-Rhazya community and 2.1b Acacia-Dodonea-Cymbopogon community. Both the sub communities differentiate from one another due to the presence or absence of Rhazya in one or the other.

Acacia-Dodonea-Rhazya Community: This community consists of 13 quadrats out of total of 40. The presence of Rhazya stricta in these quadrats distinguishes it from the other sub-community.

Acacia-Dodonea-Cymbopogon Community: This community consists of 27 quadrats out of total of 40. The presence of Cymbopogon jwarancusa and absence of Rhazya stricta is the only feature to differentiate it from the other.

Monotheca-Prosopis-Dodonea Community

This major group consists of only 35 quadrats out of total of 75. This group is formed due to the dominance of Monotheca buxifolia. It is further subdivide into 2 sub communities.

2.2a: Dodonea-Cymbopogon-Monotheca community.

2.2b: Adhatoda-Cymbopogon-Monotheca Community.

2.2a Dodonea-Cymbopogon-Monotheca community.

This sub community consists of 27 quadrats comprising of commonly occurring Monotheca and Cymbopogon species in addition to the presence of Dodonea viscosa, which differentiates it from the other sub-community. Most of the quadrats get separated in this group, thus forming the representative community of the area.

2.2b Adhatoda-Cymbopogon-Monotheca Community.

It consists of only 8 quadrats out of total of 35. Dodonea was found absent in these quadrats while Adhatoda which demarcates it from the other sub-community, forms high cover.

Use of Indigenous Plants by People

Fuel wood and Fodder : Most common plants have economic, cultural or medicinal value to the people of study area, and many species have several uses. A common combination of uses for tree is foliage as browse for animals, lopped thorny branches for fencing, and wood for fuel, tools and minor construction. All major grass species are important as forage for domestic livestock.

Gums, oils, seeds and fruits of many species are also utilized either as a food source or for medicinal preparations. Many plants are used commercially to make mats, baskets and other products.

Methods

To determine the contribution of forest and rangelands to the rural economy, dependency and exploitation of these natural resources, a study was conducted in the villages in which people belonging to different walks of life were asked for different parameters mentioned in the coming paragraphs. To find out this the socioeconomic conditions of the people living in project area were also assessed. Details of human and cattle populations and information from each family for their income resources were also sampled.

A three-part questionnaire was prepared to collect the data. The first part dealt with questions on fodder, fodder types, collection, purchase and storage.

Second part dealt with the fuelwood collection. This part of questionnaire dealt with questions on land and other resources. Fuel use was determined based on recall over four seasons and specified by kind and source. Separate questions were posed regarding the collection site with details of who collected, where they went for collection, quantity per trip, number of trips, alternative sources and time to get to the source of fuel as well as the time to collect.

The third part was developed to study the medicinal base of plants, in which elders, and hakeems (traditional healers) were visited. Native plants were explored for their indigenous medicinal importance using ethnobotanical tools of informal meetings, probing and scrutinizing.

During the course of information gathering villagers were interviewed and data on the above mentioned items were collected.

Results

Fuelwood and fodder: Pakistan with a 5000-year history of civilization is the ninth most populous country, which is highly dependent on the natural resources. It is a land of great climatic variation and is well endowed with biological resources, though many of the country's ecosystems are degraded and its rare wild species are threatened with extinction.

The actual withdrawals from the forests are also much higher than the carrying capacity of the forests. This is resulting in continuous loss of vegetation and tree cover, leading to further strain on the resource base. However, the forests of the project area provide a variety of goods and services for meeting the needs for fuel, fodder, fiber, medicine, timber and other non wood forest products, in addition to the ecological requirements of the area.

Cattle are a very logical investment for the residents of the study area, as they are for the small-scale farmers, they provide milk and meat and they are easily traded for cash or merchandize. People are dependent on fodder to feed their livestock. Fodder is a serious problem for the people living in the villages and it gets further troublesome as most of the area is more or less completely rain fed and arid. Fodder obtained from agricultural fields comprises only a small portion of the total fodder requirement. Major fodder requirements are met from grasses, which are obtained from field boundaries on their agricultural land. In addition to it, the community members also visit the communal land to collect fodder for their animals.

As far as grazing, fuel wood collection and fodder is concerned, study area is divided into three categories i.e. agricultural land, rangeland and communal land as explained by the sampled population. Most of the land type exploited for the daily use requirements in study area is communal land, which is the most over burdened by grazing, grass cutting, and fire wood collection.

Different Land types used for grazing, fuel wood and fodder collection of study area Our study results report that there is a greater participation of women as compared to men for fuel collection. Similarly females of the area collect fodder from the agricultural land, which is done in the form of weeding when crop is standing or they cut the fodder at the periphery of fields.

Grass cutting is not so damaging to the vegetation and biodiversity of the range, however, wildlife suffers due to disturbance to their habitats and possible destruction of eggs and nesting of ground nesting birds during the process.

Woodcutting for construction and fuel purposes is a basic necessity of the community members. The area containing reserve forest has been restricted and even the prohibitions could not rule out these activities completely.

In total five union councils were visited namely Soodal, Lachi, Surgul, Shakardara and Darmalak. A total of 100 samples were taken to find out whether the people collected fodder from their land which they themselves possessed or they partly purchased the fodder or they had no animal. For convenience the community was divided into following four groups:

No Animal rearing

Fodder collection group

Collection plus purchase group and

Purchase group

Statistical analysis showed that the people of study area mostly own goats and sheep with higher percentages of 37 and 26 with high standard deviation of 4.04 and 3.60 respectively. The Table-32 indicates that among domestic animals people own least number of donkeys and camels, which are mostly used as carriage source for fuel wood, fodder and drinking water.

There were 22 % people who had no animals in their houses; therefore, they did not collect any fodder. Only 8 % of the sampled people fed the animals just by purchasing fodder, and did not go for collection. More than 70 % people collected fodder from their own land, communal land or rangeland.

For the collection of wood as fuel 32 % people responded that they do not collect wood from the nearby forests and just purchase all of it from the market. It is interesting to note that all the community members that buy wood are those who fall in the "Purchase" group or "No Animal Rearing" group. Almost 60% of the people that collect wood are strict collectors and don't buy fuel wood at all.

The rest collect as well as purchase it from the market especially in winter when consumption rises due to falling temperature. In the sampled area more than 50 % peoples used gas cylinders as fuel source along with the wood.

A small quantity of kerosene oil is consumed as fuel source in main villages. The trend of using gas cylinders was observed maximum in Shakardara town and Malgin. Whereas, in Darmalak and Chechana the use of cylinders is limited. Dung cakes are being used as fuel in all those houses where cows and buffaloes are being reared. Just a couple of people buy dung cakes otherwise they are either self sufficient in it, or borrow from others or absolutely don't use it as source of energy. Figure-35 shows that it is wood, which is being consumed for fuel in the largest amount. This is because of the poor living standard of most of the population of study area and their inaccessibility to the other resources of modern world. Thus this natural resource has been put to exhaust at an alarming rate and many areas are now devoid of any tree cover due to unsustainable harvest (Chaghtai 1976). This practice has further proved to be responsible for many other disastrous effects in which one special case is erosion in study area. The other sources being used for the energy requirement in descending order are cylinders, kerosene oil, dung cakes and electricity.

For the collection of fodder and fuel wood women were seen often to join in small groups to the near by agricultural land, range land or the communal land. Women did assert that often men perform the duty of collecting wood for house hold use or at least join them on the way to forest but no such incidence was witnessed. Women alone in-groups of 6-12 were seen always carrying wood head loads. Each head load weight ranged at least between 10 and 20 kg. Commercial fuelwood cutters mostly use donkey cart or camels and in this case it was the men who did this job. Woodcutting in the reserved forest though is strictly prohibited still community members were found not only being accused of cutting wood but also admitted it explicitly.

For collecting fodder women form groups of 2-4 and move in their own agricultural land or the neighboring fields but the communal land is seldom visited for fodder collection. The house where fodder is required in small amounts for they possess goats only, the collection of fodder is not a common practice. The goats etc. are left to graze freely all over the day and for the night a bunch of grass is purchased for Rs. 2/-per day per animal. However, the people who keep cows, buffaloes and donkeys have to buy fodder in larger quantities. If green fodder is collected from the fields, even then the dry fodder like oil cakes (khal) and wheat straw has to be purchased. The average expenses on a buffalo are Rs. 50-60/day but the cows consume less.

It is very clear from Table 33 that grasses are highly consumed fodder (50%) if available. Agriculture crop residues are next to the grasses as fodder while trees were third to grasses.

Most of the people in the study area use open lands as pastures to graze their animals. These pastures are either part of their own land, community land or some times even government land. All the animals are left in the grazing land for the whole day, which includes goats, sheep, cattle, camel and donkeys. Goats and camels cause more destruction to the vegetation of that area as compared to the animals because they eat away any available palatable species and such condition was observed to be highly severe and intense in Shakardara area (Census Unit 1) where capacity of the land to produce fodder is less because of low rain fall and rugged topography but there are large number of animals which are being reared due to that degradation of land and vegetation was frequently observed.

As far as grazing, a trend though rarely observed but is interesting to note that people send their animals to the communal land along with a flock and the shepherd charges Rs. 30/-per goat per month and Rs. 50/-per cow hence to take animals for grazing is also observed as profession.

The rapidly increasing human and cattle population in study area is putting immense pressure on the natural resources. There are continuous attempts to put more and more areas under plough and conversion of forestlands and grasslands to agricultural areas has not only disrupted the biodiversity, natural ecosystem, ecological balance in many areas, but it has also caused fragmentation of the natural habitat, disrupting the movement of wild life and encroachment of agricultural land has reduced the forest cover area.

Apart from all the discussion above this is a fact that whether the fuel wood or fodder is collected or purchased depends on three major factors. They are:

Financial status or the income group; the collection practice is observed very commonly because a large number of community cannot afford to buy food for their children as well as their animals due to meager income from various sources. Most of the area is rain fed, therefore, the scarcity of water, insufficient wild grasses and inappropriate conditions to grow the cultivated fodder, all lead them to purchase fodder.

Season; the grasses are available from July onwards till winter turns very severe. In spring and summer green fodder is so scarce and limited that people are compelled to buy fodder for their animals.

The fuel expenses of one household that consist of an average of 10 people are approx. Rs. 500/ per month. This includes all the oil, gas and fuel wood expenses. It entirely depends upon the possession of agricultural land, availability of water and the financial status of the family that what percentage of this expense is saved by their visits to the communal land. They may save it all just by restricting themselves to collection of fuel food and no use of kerosene oil and natural gas at all, as observed in 60% of the houses. While on the other extreme all the fuel whether in the form of wood or gas is purchased but the number of such people was very little and it was only in 5% of the families who were doing so. The rest of the families fall in between the two extremes partially by collecting/ purchasing fuel wood and purchasing oil and gas.

Acacia modesta, Acacia nilotica and Dodonea viscosa are the most favorite fuel wood species. Dalbergia sissoo is expensive and rare and is used for construction purposes but not for burning. Cymbopogan jwarancusa, Heteropogon contortus, Aristida sp, Poa annua, Cynodon dactylon, Chloris sp etc. are commonly growing grasses in the area and are used as fodder.

Before the new grasses have sprouted in early spring, people rely mostly on tree fodder from the forest and communal lands. Herbaceous plants of the spring flora in the study area (and some tree fodder) form the main fodder resource. Grasses are harvested from the edges of the agricultural fields and from communal land areas of Guzara Forest from September to October. The palatable species wherever available in smaller or greater quantities grazed through out the whole year. When food available in the field is not sufficient, then it is supplemented by the stored fodder. The stored fodder comprises of fodder grasses together with maize and wheat stalk which is gathered after the grain, has been harvested and some time also purchased to meet the needs. This stored fodder is also used during winter season.

Species of Special Significance

A species of dwarf palm, locally called Mazri (Nannorrhops ritchieana) is used commercially to make mats, thatches and ropes. The plant is more common in Balochistan and NWFP. In the study area only deformed plants were found in very small patches. The poor health of the plant could be due to overuse by local communities in the past, which has diminished its population. Leaves are important medicinally and are used against diarrhea, dysentery and also purgative in case of cattle.

Olea ferruginea was once present through out the study area as an indigenous plant species. Now its presence has shrunk to the high ridges because of its extensive and lavish cutting. This plant has vanished from many of the places of the study area, i.e. Rakh Sobar, Rakh Topi etc. As the whole of the plant is important in one way or other, that is why this plant is under great stress. Wood is considered to be very hard and used in making the handles of agriculture tools. Walking sticks are also formed from its wood. Its wood is sold in the market in the shape of sticks having an average diameter of 4 inch and 4 feet in length. The decoction of leaves is used for toothache. Leaves are bitter, astringent and antiseptic. The leaves are boiled and used for hoarseness of voice. Fruit is known as antidiabetic. Root is useful for asthma.

Fauna

Study area contains resident populations of many animals. In addition, some migratory birds pass through study area on their way to and from summer and winter feeding grounds, while other use the area as wintering area. Many species vary in activity patterns between one time of the year and another. This report deals with only resident species of the study area. Field investigations were carried out in August and September 2001.

Habitat Types

Habitat types in the area correspond with its various landform features. The area consists of vast flat plains traversed by several dry streams originating from low hills. In all, there are seven main habitat types, each supporting different communities, which have been discussed in the vegetation portion.

Methods

Choice of Sites

It was important for the purposes of the survey that fauna be investigated throughout the area in a relatively unbiased way. Initial site selection was carried as for flora.

Sampling sites were partly randomly chosen through the study area, using a stratified random sampling technique, however, some samples were taken for other reasons, such as documenting changes down a topographical gradient, or ensuring that all major habitat types were sufficiently studied. Faunal surveys were then collected as a subset of those points to make possible subsequent correlation between the fauna and flora sets of data.

Sampling Techniques

Because many vertebrates are inconspicuous, nocturnal, or avoid detection in other ways, a variety of techniques were used to establish the presence of a species for quantifying wild mammalian herbivore populations. These techniques include capture/ mark, indices counts, random and stratified sampling and total counts, etc. The suitability of a specific technique for a given area depends upon several factors that include:

The innate behavior of animal species;

The nature of terrain;

The time available;

Keeping in mind all above-mentioned factors, it was concluded that the only practicable and reliable survey technique would be a total count or census for large ungulates e.g. urial and chinkara. The following techniques were used in surveying the fauna of study area.

Sighting with the help of binoculars (Nikon 7x50);

Collection of pellets and other signs;

Recordings calls;

Literature reviews; and

Personal communications

Population surveys of urial and chinkara are notoriously difficult because of the high motility of the species, cryptic coloration and shy nature i. e., they can detect and flee away from humans at a great distance. All surveys were conducted from the ground while hiking and urials were spotted from observation points and along ridgeline travel routes. Drop-off points and observation points were documented using Global Positioning System (GPS) technology.

The area was surveyed as per accessibility and travel paths with a view to making the sample representative of all micro habitats found in the area quantitatively. The area was surveyed approximately for six hours daily on the average to record observations. Animals were observed by walking along ridges and ravines.

The project area was divided in to five blocks and observations on animals were made in each block by scanning ridges and ravines. Time, date and location were recorded for each observation along with species, age and sex class.

Field investigations were carried out in the study area to ascertain the presence, distribution and density of wildlife species especially urial and chinkara in the area. Reliable information on the wildlife of the project area with regard to species, their distribution and status, is an essential requirement for further management.

Avi-fauna

It is important to monitor bird populations for two main reasons. First, birds form an important and popular wildlife resource, which is worth conserving in its own right. Second birds provide valuable indicators of the state of the environment. They are usually high in food chains and so particularly susceptible to environmental changes. The point counts and line transect techniques were applied for quantitatively sampling bird populations. In all, a checklist of 57 bird species was compiled for the area on the basis of field observations made during the surveys.

Passerine birds were mostly observed chirping in the vicinity of water or in the agricultural fields. Grey partridge calls were frequently heard in the early hours and late afternoon. Sandgrouse was encountered on the tracks. Red turtledove population was dominating through out the study area. Purple sunbird with brilliant plumage and long, de-curved bills for probing flowers for nectar were observed through out the area. Birds of prey and Scavenger birds were observed gliding on the top of mountains. Kestrel was observed along the narrow cliffs and also in the open valley hovering high up in search of food.

Occurrence of major game bird species, viz., chukar, see-see, black and grey partridges makes the area all the more important from biodiversity conservation point of view. The populations of these species are prone to illegal hunting and capturing round the year.

6

Forest Cover in India

INTRODUCTION

India's National Forest Policy, 1988 aims at maintaining 33 percent of country's geographical area under forest and tree cover. FSI, since the 1980's has been making assessment of forest cover in the country, using data from remote sensing satellites, on a two-year cycle and publishing it in the State of Forest Reports (SFR). However, assessment of tree cover, i.e. trees that were not captured by the satellite sensor, was never made. Thus, a complete picture of forest and tree cover in the country that could be compared with the goal of 33 percent set for the nation was not available. The present assessment of forest cover, carried out by digital processing of satellite data, can encompass forests and tree crops having 10 percent or more canopy density and with area more than 1.0 ha.

Due to technological limitations it is not possible to provide information of tree patches, which are less than 1.0 ha in extent. However, small patches of trees in plantations and woodlots, or scattered trees on farms, homesteads and urban areas, or trees along linear features, such as roads, canals, bunds, etc. constitute substantial tree wealth of the country. The tree cover comprising such trees would make significant contribution to the over all forest and tree cover of the country. For SFR 2001, the tree cover in the country has been assessed using technique of stratification, sampling, field inventory, data processing and statistical analysis. This chapter gives method used and steps involved in assessment of tree cover and presents the results for the country, different strata or zones and all the States/UTs.

TREES OUTSIDE FORESTS AND TREE COVER

FSI has been conducting inventory of trees in non-forest areas or Trees Outside Forests (TOF) for the past about 10 years. Inventory of TOF for a state is done by taking rural non-forest area of a state as population. The state is stratified according to agro-ecological zones and each zone is further stratified according to districts. Villages within a district are considered as sampling units. The villages are selected randomly while the sample size is such that it provides estimates within 15 percent permissible error (higher levels of accuracy increases the sample size substantially making the inventory work more costly and time consuming). An accuracy level of 85 percent appears statistically sound and satisfactory for such assessments. Inventory data is then collected on pre-designed formats. The trees are classified into eight categories viz. farm forestry, village woodlots, block plantations, road side, pond side, railway side, canal side and others. Data is processed and tree cover due to TOF is estimated by using ratio estimation method.

Using this methodology, FSI has inventoried about 180 districts in 22 states of the country so far. It may be noted that all trees growing outside recorded forest area constitute TOF, but for tree cover, only those trees in non-forest areas have to be considered that do not constitute block plantations or woodlots more than 1 ha in area since these have been included as forest cover. Moreover, TOF assessment by FSI so far was being done at state level. In order to assess tree cover at the national level, a somewhat different method of stratification and sampling design had to be devised.

STRATIFICATION OF COUNTRY INTO PHYSIOGRAPHIC ZONES

Whenever any feature or a variable is to be estimated in a large population, stratification of population into reasonable number of homogenous groups or strata is a standard statistical technique. This procedure optimises the sample size, increases accuracy of assessment and reduces the overall cost of estimation. For assessment of countrywide tree cover, the country has to be stratified into geographical zones within which the tree species, density and growth etc. are more or less comparable. Since, besides forest trees, we are dealing with trees largely grown outside natural forests, mostly consisting of fast growing exotic trees or fruit trees or multi-purpose trees, stratification of country simply according to forest types will not be suitable. Similarly, stratification into bio-geographical zones as done by Wildlife Institute of India, Dehradun for wildlife habitats is also not entirely satisfactory.

However, these classifications will have a bearing on the boundaries of strata finally selected for tree cover assessment.

A workshop and a follow up meeting on Global Forest Survey were organized by FSI in collaboration with FAO wherein it was suggested and accepted that physiographic zones can be used as strata in assessment of tree resources in the country. A physiographic zone, on the basis of topography, latitude and altitude, besides climatic and soil properties, constitutes geographical areas that exhibit broad similarities in factors responsible for the growth of tree vegetation. The National Atlas and Thematic Mapping Organisation (NATMO), Kolkata in its Atlas of Forest Resources in India (1976) has classified India into seven physiographic zones namely, Northern Mountains, Indo-Gangetic Plains, Central Highland, Deccan Plateau, East Coast, West Coast and Islands.

In order to improve the quality of inventory data, these zones were subdivided and/or modified further on the basis of homogeneity of tree vegetation observed in different parts of each zone. For example, the tree vegetation composition of Western and Central Himalayas is different from Eastern Himalayas and accordingly the physiographic zone covering Northern Mountain was divided into two zones namely, Western Himalayas and Eastern Himalayas. Similarly, Nagaland, Manipur, parts of Assam, Meghalaya, Tripura and Mizoram, included in the Nothern Mountains above, support similar tree vegetation which is different from that supported by Eastern

Himalayas or Eastern Plains and, therefore, has been put under a different physiographic zone, namely North East.

The vast geographical area under Indo Gangetic Plains was sub-divided into three physiographic zones namely, Western Plains, Northern Plains and Eastern Plains. The Kutchh and Surendranagar region of Gujarat (included in the West Coast as per NATMO), has tree vegetation composition similar to Western Plains and, therefore, this region was included in Western Plains. Likewise, the Brahmaputra Valley in the north-east which was hitherto included in Northern Mountains zone supports tree vegetation similar to Eastern Plains and therefore, has been included in the Eastern Plains.

Following similar reasoning, the Deccan Plateau, where tree vegetation is highly varied, was sub-divided into five zones namely, North Deccan, East Deccan, South Deccan, Western Ghats and Eastern Ghats. The islands of Arabian Sea and Bay of Bengal have been clubbed with West Coast and East Coast, respectively. In this manner, the whole country has been classified into fourteen physiographic zones:

1. Western Himalayas (WH)
2. Eastern Himalayas (EH)
3. North East (NE)
4. Northern Plains (NP)
5. Eastern Plains (EP)
6. Western Plains (WP)
7. Central Highlands (CH)
8. North Deccan (ND)
9. East Deccan (ED)
10. South Deccan (SD)
11. Western Ghats (WG)
12. Eastern Ghats (EG)
13. West Coast (WC)
14. East Coast (EC)

The list of districts falling within each physiographic zone, completely or partially, has been given in Annexure-III for information.

SAMPLING AND DATA PROCESSING

For the purpose of this study data was collected from each physiographic zone so that estimates of tree cover could be generated within 25 percent permissible error. The permissible error was kept on the higher side, as this was the first attempt to assess tree cover at national level with limited data or sample size. Lower permissible error would have required much larger sample size. In subsequent years, as data volume grows, the assessment would be more accurate. Accordingly, required number of sampling units, i.e. villages, was enumerated employing random sampling within physiographic zones. For the present study, data of 1,154 villages from 123 districts spread over the entire country and covering all physiographic zones were used.

The above mentioned data pertained to rural areas only. The tree cover scenario in urban areas is apparently different and the tree wealth of urban areas also needed to be estimated separately. In this context a methodology for assessment of urban non-forest area was developed and tested in the field. The definition of urban area has been taken from Census of India that identifies urban areas according to population. The towns have been classified into five categories on the basis of their population. These categories have been used as criterion for their stratification for the present study.

The sampling frame used for inventory was taken from the National Sample Survey Organization (NSSO). NSSO conducts Urban Frame Survey (UFS) in urban areas to form blocks (called UFS blocks) on the basis of population (600-800 persons per block) or households (120 to 160 households per block). Consequently, UFS block maps with identification particulars are prepared by NSSO.

For inventory of urban areas, stratified random sampling method was followed by taking categories of town as strata and

UFS blocks as sampling units. About 2.5 percent UFS blocks (or a minimum 4 UFS blocks) were selected randomly in each category of town (strata) for survey and data on pre-designed formats were recorded. Data of 281 UFS blocks from 15 districts covering all physiographic zones were used for assessment of tree cover in the urban areas.

AGGREGATION OF SAMPLE DATA

The data obtained from sampling units have to be aggregated over the corresponding stratum using ratio method of estimation. At this stage care has been taken to see that aggregation of sample data does not result in any overestimation yielding inflated or erroneous values. The non-forest area within a stratum over which the aggregation has to be done may contain certain unculturable lands that cannot support tree vegetation. Such lands include wetlands, riverbeds and perennial snow covered mountains. Also, the extent of area under forest cover outside forest area, consisting of plantations and woodlots more than 1 ha in area, is not to be included for the tree cover assessment. All these areas must be subtracted from the non-forest area. This would provide us with an entity defined as "Culturable Non-forest Area" (CNFA) in this report. This is the net area over which the sample data of tree cover can be aggregated to obtain the estimates for tree cover.

For this purpose, data on the extent and distribution of all components of unculturable non-forest area would be needed. Information about wetland and water bodies was taken from the Project Report "Wetlands of India" conducted by Space Application Centre (ISRO), Ahemdabad, June, 1998. The data on area under rivers and river beds was obtained from "Problems and Prospects of National Resources Management in Indian Himalayas" by Samra & Dhyani, National Research Centre on Cold Water Fisheries-Vision 2020 and "National Water Resources at a Glance" from website of Ministry of Water Resources.

The information about cold desert and snow was taken from the paper "Satellite Mapping of Alpine Pastures in the Himalayas" by J.B. Lal et.al.; International Journal of Remote

Sensing (1991, vol. 12, No. 3). These areas were proportionately divided among recorded forest, rural and urban areas within each physiogrpahic zone. The areas assigned to rural and urban areas were considered as unculturable non-forest area and deducted accordingly.

Besides above mentioned unculturable areas, the area under forest cover outside the recorded forest area (i.e., blocks of trees greater than 1 ha) also had to be subtracted from non-forest area to arrive at an accurate estimate of CNFA. These areas were estimated by using sample inventory data obtained from various districts spread over all the physiographic zones. These estimates were also compared with forest cover of State and UTs and necessary correction factors were applied, especially for those states where the forest cover was more than the forest area. The areas thus estimated above were also subtracted from the non-forest area to assess Culturable Non-forest Area (CNFA). Physiographic Zone wise and State/UT wise estimates for the components of CNFA are given in Tables 4.01 and 4.02, respectively.

These tables reveal that only 68.3 percent or about two thirds of the total geographic area of the country is culturable non-forest area. The individual components of forest area and unculturable areas as estimated here (with respect to country's geographic area) are the recorded forest area (23.4%), blocks of forest cover (>1 ha) outside forest area (3.2%), wetlands & water bodies (0.7%), rivers & riverbeds (0.4%) and cold deserts & snow bound areas (4.0%).

RELATIONSHIP BETWEEN DIAMETER AT BREAST HEIGHT (DBH) AND CROWN AREA

The inventory data, consisting of tree species and dbh, collected from sampling units in rural and urban areas within a zone gave us an estimate of average number of trees per hectare in that zone. Also, the total number of trees of different species falling in different diameter classes (10-20cm, 20-30cm, 30-40cm, 40-50cm and >50cm) in each physiographic zone was computed. In order to convert the number of trees into tree cover, relationships had to be developed between diameter of tree and its crown area for each species within each zone. Also, it has to be decided at what canopy density should tree cover be estimated.

During inventory, species and dbh of each tree measured were recorded. The tree species which constituted more than 1 percent of the total trees in a physiographic zone were identified as prominent species. A sample of at least 30 trees for each of such species, representing all the diameter classes, was selected. The dbh and crown width of selected trees were recorded and the area of the crown of each tree was calculated. A regression model between dbh (as independent variable) and crown area (as dependent variable) was developed for each prominent species in each physiographic zone. Where a tree species was distributed over several physiographic zones, data of adjoining zones were clubbed together to harmonise the crown area and diameter relationship. Using the model, crown area for each diameter class (at the mid point of each diameter class) was estimated.

CANOPY DENSITY AND TREE COVER

The area under tree cover is a "notional" area. It is an area covered if all the trees included in the assessment of tree cover are brought together to constitute a block of tree land or forest. However, the extent of this notional area will depend on the canopy density at which the trees are presumed to be grouped together.

What canopy density should be chosen to assess the tree cover? In order to avoid subjectivity in deciding this important parameter, it is necessary to examine possible options and to take decision on the basis of scientific reasoning and sound logic.

The several possible options available are mentioned and discussed below:

(i) The canopy density of forests included in forest cover range from 10 percent to 100 percent. The density at the middle of the range, i.e., 55 percent, can be one option for canopy density of tree cover.

(ii) The average density of existing forest cover can be another option. The dense and open forest cover, estimated in Chapter 2, when given weights of 0.70 and 0.25, respectively (being the mid points of density classes for dense and open forest cover), the average density of country's forest cover works out to be 52.8 percent.

(iii) The cut off point between dense and open forest cover, i.e., 40 percent, can also be an option.

(iv) The tree cover is generally composed of trees, plantations, linear tree crops and wood lots in private or institutional lands. Due to intensive management and better protection, the density of such tree crops is usually found to be dense.

Theoretically, the maximum crown density for an even-aged monoculture crop without any under-storey is about 80 percent. Keeping a small margin for gaps and mortality in plantations, an average density of 70 percent can be presumed in general. 70 percent is also the mid point of the range of density (40 to 100 percent) used for the class of dense forest cover. Thus, density of 70 percent can be another option to be used for tree cover.

Of the four options mentioned above, the fourth one appears to be the most reasonable and justified.

It is also the most conservative option. It is quite practical and safe to select a conservative canopy density for tree cover assessment. By selecting any other option, i.e., a lower density for assessing tree cover, one is open to allegation that the extent of tree cover has been inflated artificially.

Instead, the tree cover in the country should be increased by actually planting more trees on wastelands, agriculture farms, linear strips, homesteads and unused lands available in rural and urban areas. Besides forest departments, concerted efforts in this direction are required form departments of horticulture, agriculture, industries, etc. and from municipalities, urban development authorities, various institutions and tree growing entrepreneurs.

Having decided 70 percent as the canopy density for tree cover, the numbers of trees have to be converted into tree cover by using the relationships between tree diameter and crown area. The number of trees whose crown can cover 7,000 m^2 would constitute one hectare of tree cover. The estimates generated from dbh and crown area relationships for prominent tree species in each physiographic zone

ASSESSMENT OF TREE COVER

Once the species wise and diameter wise distribution of trees in culturable non-forest area in each physiographic zone has been estimated and the number of trees required for one ha tree cover for each diameter class has been calculated, the tree cover within a zone and the country can be easily assessed. Serially adding tree cover for all diameter classes over a species and then adding over for all species separately for rural and urban areas in a zone would give the tree cover for that physiographic zone. Finally, adding tree cover for all the physiographic zones would yield the estimated tree cover of the country. The total tree cover for the country has been estimated as 81,472 km^2 or about 2.48 percent of the country's geographical area. This estimate, as per the accuracy of tree cover assessment computed and described in Annexure-II, has a standard error of 1.83 percent.

TREE COVER IN THE COUNTRY: PHYSIOGRAPHIC ZONE WISE

The estimates of tree cover for each physiographic zone is given in Table 4.03. It is noted that South Deccan physiographic zone has maximum tree cover of 11,468 km^2 that is 3.92 percent of its geographic area, followed by East Coast having 10,419 km^2 (6.22 percent) and Northern Plains with 10,098 km^2 (3.28 percent). The area under tree cover in any region largely depends on the extent of CNFA and the density (i.e. the average number of trees per hectare). It is noted that the density of trees in the culturable non-forest areas is maximum in Western Ghats (23.5 trees/ha) followed by East Coast (19.7 trees/ha), West Coast (18.2 trees/ha) and Northern Plains (15.6 trees/ha).

Percentage of tree cover with respect to CNFA is another important parameter. It indicates the actual potential existing in different zones for increasing the extent of tree cover. This percentage is highest for Western Ghats followed by East Coast. On the other hand, this proportion is quite low for Western Plains and Eastern Ghats where there appears to be large scope for tree planting activities outside forest areas. However, the information on tree cover has to be seen in combination with the extent of forest cover outside recorded forest areas.

The standard error of estimated tree cover falls between 5.87 percent (for Northern Plains) and 22.47 percent.

TREE COVER IN THE STATES AND UNION TERRITORIES

Tree cover data was processed further to provide information of tree cover for each state and union territory. One state may fall in one or many physiographic zones; accordingly estimation procedures were used to develop state level estimates. As such, area of different physiographic zones within one state is considered as separate strata. CNFA corresponding to the State/ UT has been ascertained in a similar fashion as in case of physiographic zone. CNFA corresponding to different physiographic zone falling within a State/UT was also ascertained. Using estimates of trees/ha of different physiographic zone, estimates of total number of trees for the respective state was calculated following stratified random sampling estimation procedure. However, it may be noted that the State/ UT wise estimates for tree cover are only indicative in nature and may have lower levels of accuracy since the sample size was calculated only to provide estimates at the physiographic zone level.

Tree cover constitutes largest area in Andhra Pradesh (9,011 km^2) followed by Maharashtra (8,269 km^2), Uttar Pradesh (7,545 km^2) and Karnataka (7,446 km^2). Considering the percent of geographical area of the State/UT under tree cover, the highest rank goes to Pondicherry (7.19 percent) followed by Dadra & Nagar Haveli (5.54 percent), Tamil Nadu (4.65 percent), Bihar (3.92 percent), Karnataka (3.88 percent) and West Bengal (3.68 percent).

Considering the percentage of tree cover with respect to CNFA, which indicates the actual potential different States/ UTs have for increasing area under tree cover, it is found that Lakshdweep, Dadra & Nagar Haveli, Pondicherry and Andaman & Nicobar Islands cover more than 7 percent of their CNFA's. The states of Tamil Nadu, Bihar, Karnataka, and Kerala follow with over 5 percent CNFA under tree cover. Analysing these figures in combination with the extent of forest cover outside recorded forest areas in different States/UTs, the concerned governments can determine where and how much scope is there to enhance tree-growing activities.

Concluding Remarks

The assessment of tree cover for the whole country requires a huge volume of latest data. In the present assessment, the sample size, especially for survey of tree cover in urban areas, was quite small. In the future assessments, endeavour would be made to increase sampling intensity in urban areas and in those physiographic zones where the estimated error is over 15 percent. Information on State and UT wise tree cover is more meaningful for planning purposes. Sufficient data would be collected from each State/UT to compute such information with greater accuracy.

7

Arid Zone Agroforestry: Dimensions and Directions for Sustainable Livelihoods

INTRODUCTION

Drylands cover more than 60% of the earth's surface. Arid zones are described as a part of the drylands, and have the most severe edapho-climatic conditions. The major distinguishing feature for defining and for planning for the arid zone is the low rainfall (below 500mm or Aridity Index <0.20) with more than 50% inter-annual variability. This makes a great difference in terms of the nature of the ecosystem, the socio-economic environment and the challenges for sustainability. High wind and solar regimes further increase the effect of rainfall variability and the whole complex makes a fragile ecosystem in which small disturbances may cause great loss to the sustainability, which are sometimes irreversible (Houerou, 1996). The hot arid zones of the world are economically and environmentally disadvantaged, with unique problems.

The ecosystems of these zones are highly fragile with large risks that cause severe impediments to development programs. Of the total land area in the world, arid zones cover 18.8%. The hyper-arid zones account for 22.3% of the total arid zones (FAO, 1989). These arid zones are diverse in terms of climate, soils, vegetation, animals, and the lifestyles and activities of the people. Little but variability in rainfall and the presence of distinctive periods of drought are the characteristics of the arid tropics. Often, the terms drought and aridity are used incorrectly. A drought is a departure from average or normal conditions in which shortage of water adversely impacts on the functioning of ecosystems, and the resident population of people, whereas aridity refers to the average conditions of limited rainfall and water supplies, not to the departures there from. In general, arid zones are characterized by pastoralism and little farming but there are always exceptions (Malhotra, 1984).

Arid zones are found in almost all the continents of the world. In North America, arid zones are found in the southern US and Mexico, and in South America they are located in Brazil, Argentina, Chile, and in some other countries. In Africa, substantial parts of Saharan Africa, Ethiopia, and Namibia, are classed as arid environments. The hot arid regions of India lie between 24° and 29° N latitude, and 70° and 76° E longitude, covering an area of 31.70 million hectares, and involving seven states: Rajasthan, Gujarat, Punjab, Haryana, Andhra Pradesh, Karnataka, and Maharashtra. In total, 11.8% of the country is under a hot arid environment. The arid regions of Rajasthan, Gujarat, Punjab, and Haryana together constitute the Great Indian Desert, better known as the Thar Desert, which accounts for 89.6% of the total hot arid regions of India, and constitutes the principal hot arid zone of the country.

Table : Distribution of arid regions in different states of India.

State(s)	Area (10^6 hectares)	Percent of total
Rajasthan	19.61	61.0
Gujarat	06.22	19.6
Punjab and Haryana	02.73	09.0
Andhra Pradesh	02.15	07.0
Karnataka	00.86	03.0
Maharastra	00.13	0.4
Total	31.70	100

In arid zones, vegetation is typically sparse, and is comprised of perennial and annual grasses, other herbaceous plants, shrubs and small trees. The native plant species have adaptations that enable them to reproduce, grow and survive in the most inhospitable edapho-climatic conditions. Some plants have evolved special root systems, while others have unique leaf characteristics that allow them to withstand prolonged periods of drought. The number of tree species is very limited in arid zones, and in general, they are very slow growing due to limitations of environmental conditions, but nowhere in the world are they are so intricately associated with the life of human beings.

To evade or minimize the adverse affects of frequent droughts, the native peoples in arid zones have often developed production systems in which woody perennials have a very important role, both from a productivity as well as a resource conservation point of view (Tewari, 2000). Their centuries-old experiences, under diverse socioeconomic and cultural conditions, were passed from generation to generation and people have established well-contained systems of production (silivipastoral/agrisilvicultural/agrisilvipastoral/livestock husbandry) that accrue maximum benefits from their woody components. When arid zones have been looked at from the perspective of forestry development, the focus has been on the management of trees and shrubs that are either native to a particular arid zone, or have been introduced, especially for conservation purposes.

However, concern for trees and forests has increased in arid zones as it has in many other regions in recent years. This concern has been fuelled mostly by the disappearance of large areas of woodlands and trees from such fragile ecosystems in many countries of world. In the arid zone, closed forests are seldom available. In India, traditional animal husbandry and agroforestry practices have been used to manage parklands, rangelands, reserved silvipastures near holy places, lay farming, and run-off farming (traditional watershed management). Trees are managed mainly for their non-wood forest products (NWFP) and their environmental services. Animals are an essential part of the production system. The whole system has been traditionally developed to spread the risk of drought in diversified components and efficient utilization of scarcely available natural resources.

ARID ZONE AGROFORESTRY

As stated above, the prevalent form of forest management in arid lands is agroforestry, which is concerned with the management of trees and shrubs for conservation and sustainable development. Agroforestry in arid lands differs in many ways from the timber forestry practiced in moist tropical and other temperate ecosystems and in many cases its applications are broader in their scope. These applications include production of wood for fuel and poles; horticultural practices for fruit and nut production; range management for fodder and forage production; forestry to modify microclimate for improved agricultural crop production; and protection forestry on land susceptible to wind and water erosion.

Many rural people combine these applications of agroforestry with other land uses linked to their needs and social values. The value of forestry in the arid zone is difficult to evaluate in monetary terms. Arid zone agroforestry therefore needs to be understood broadly as management of trees and shrubs to improve the livelihood and quality of life for rural people in the arid environment. It is in this diverse context that policy developemnt must take place if sustainable development of forest resources in the arid zone is to be achieved.

Agroforestry is being followed traditionally in the arid zone to minimize the risk of drought and sustain life in climatic adversities. Several need-based systems have developed traditionally and some have been improved or incorporated in recent times.

TREES IN LIFE SUPPORT SYSTEMS-TRADITIONAL AGROFORESTRY

The image of the desert as little but vegetation-less, unbroken, stretches of sand and inaccessible terrain conditions is shattered upon entering the arid zones. There are sparsely distributed trees, beneath which are cultivated arable crops (especially during the kharif season, as agriculture is predominantly rain fed) and/or grasses, and other herbaceous flora, in long stretches interspersed among widely scattered village settlements and dhanis (a unique settlement pattern of the Thar intended for the life of agriculturists' families during the active cropping period away from the village, but nearby their fields). The Indian arid zone is a vegetated desert having a very rich floral diversity of about 682 species, out of which 131 arc known for their economic uses (Bhandari, 1990).

Farmers in the hot Indian arid zone grow arable crops in association with tree species. In fact, such integration of arable crops with trees in the farming systems is a unique, combined, protective-productive system that works on the principles of ecology, productivity, economics, and sustainability. These systems are now generally referred to as agroforestry. Because most trees are drought resistant, they are still able to provide fuel, fodder, fruit and other products, when the crops fail, as frequent droughts are a common phenomenon. Thus, trees have a very important place in the life of people in the arid zone of India, as they are directly related to the livelihood of inhabitants, and also provide the important service of climate moderation in an inhospitable environment (Sharma and Gupta, 1996).

RUN-OFF FARMING

This is a system of growing crops on harvested and stored water in the farm soils by constructing an earthen dam or a bund across the gentle slope of the farmland in the valley bottom. Shallow, gravelly and rocky uplands normally used for grazing are used as catchments for efficiently harvesting the runoff water. This is locally known as the Khadin system. The major components of the Khadin system are the water collecting area (low rocky catchment), contour bands (channels), moisture storage basins (khadin farm), water impounding mechanisms (bands, spillways and sluice) and a zone of cultivator's settlements.

In this region monsoon rainfall is scanty, erratic and generally the water requirement of most of the crop plants is not fully and properly met. With the result, cropping in the Kharif (rainy season) often becomes a gamble. To offset such a risk and to have some assured crop yield in such environment, one very thoughtfully developed traditional system known as Khadin, has been followed in this region for centuries (Kolavalli and Whitaker, 1996). This system makes the use of shallow gravelly and rocky uplands as the catchments for efficient harvesting of runoff water, which otherwise normally used for grazing. The water is trapped in the low lying valley plains that are converted into farmland where either the kharif or the rabi (winter) crops are raised, depending upon the amount of rainfall and consequent runoff received during the monsoon. The system takes into account not only the vagaries of the climate to raise the crop with such a low rainfall but also prevent any soil deterioration.

SILVI-PASTURE

In the arid zone there are areas where rainfall is below 200 mm/yr and with this rainfall food production is very difficult. On the other hand there are some grass species, e.g. Cenchrus cilieria, Lasiurus sindicus etc., which are very well adapted to such climate and make natural rangelands. The climax tree species like Prosopis cineraria and Zizyphus numularia naturally come up in these rangelands and make a silvipastural system. Animals like cows, goats and sheep are the part of this farming system.

Traditionally the areas near holy places like temples are kept reserved for silvipasture with very well established social fencing. This is a very good example of sustainable management of common property resources that still exist in several places. The silvipasture systems are strictly used according to the land's carrying capacity. Also these systems cover large areas i.e. 100 to 500 km^2. These systems have low productivity but are an example of an excellent synergism between the nature and the natives and that's why they have been sustained for centuries (Shankarnarayan et.al. 1987).

AGRI-SILVICULTURE

This is the most popular system in the areas receiving rainfall between 200-400 mm/yr. People protect the naturally germinated seedlings of useful trees like P. cineraria, Z. numularia, Tecomela undulata, etc., which come up in the crop fields. Mixed cropping of pearl millet, moth bean, cluster bean and sesame is carried out under these trees. These trees do not compete with the crops but rather are complimentary in terms of improving the microenvironment. These trees give fodder, fuel, fruit etc. even when crop fails during drought, thus provide sustenance to the farmer during drought. The P.cineraria and Pearl millet agroforestry is well known around the world. Some other aspects of this system are discussed with the Non wood forest products.

NEW SYSTEMS DEVELOPED

Some of the agroforestry systems have been developed after research and on farm testing are discussed below.

Agri-Horticulture. After two decades of experiments on research farms and farmer's fields, a new agri-horticulture system of jujube (Ziziphus mauritiana) inter-cropped with arid legumes e.g. clusterbean/mothbean/greengram has been developed for the areas receiving rainfall of more then 250 mm. This agri-horticulture system has been found to give better and earlier production, year-round work and resilience to erratic rainfall. In this system initially the jujube saplings are raised in the farmer's nursery by stored rainwater (farm pond) and then saplings are transplanted at a distance of 5 x 10 meters irrigated twice in first two weeks; thereafter the trees receive only harvested rainwater. Arid legume crops are raised successfully in the interspaces of jujube.

It is a short duration (50-75 days) crop, sown at the onset of monsoon and harvested in September/October, i.e. before the full growth and blooming of jujube. This offers least competition and both are complimentary as the jujube provides a favorable environment for arid legumes while legumes fix nitrogen, and leaf fall adds organic matter to the soil. Besides fruits and wood from pruning (for fuel or fencing), leaves as fodder are the additional products from jujube. Overall productivity of the system in normal and drought years shows less variation in yield due to variations in rainfall. This system can provide economic returns of Rs 28000-37000/yr as compared to the sole cropping of arid legumes (Rs. 8000-12000) with low input supply, besides non-tangible environmental improvements which give resilience to drought.

Lay farming. In this system, on one piece of land a rotation of grasses for 4-6 years, followed by food grain crops like pearl millet or legumes for 2-3 years is followed and then the land is left fallow for 2-3 years. The whole farm is divided into parts in such a way that every year all three practices, i.e. grass production, crop production and fallow, are available in one or another part.

In this traditional system the rotation was designed considering the poor carrying capacity of arid soils.For that purpose first grasses are grown which improve the soil's physical and biological conditions, then good production is obtained, followed by a fallow period for natural restoration of the soil. During the fallow period several weeds come which improve the soil and the life cycle of various pests is broken. This system is gradually vanishing because of several social changes which are based on the thinking of exploiting nature rather working in synergism with nature.

Wind breaks/shelterbelts. Wind erosion, a high thermal regime and hot desiccating winds are some of the serious problems, which affect the establishment, growth and yield of crops in arid areas. A mixture of trees and shrubs planted across the wind direction help in reducing the wind speed. Evaporative demand studies conducted at CAZRI with three row shelterbelts of Acacia tortilis, Cassia siamea and Prosopis juliflora as side rows and Albizzia lebbeck as the central row were found effective in reducing wind speed and wind erosion. Micro-shelterbelts of tall growing plants have also been found quite effective in providing protection with about a 20 to 40 per cent increase in yield of vegetable crops grown with irrigation.

CHALLENGES FOR SUSTAINABLE ARID ZONE AGROFORESTRY

Agroforestry has been followed since time immemorial for sustainable livelihood in the arid zone, but in the recent past some changes have occurred, due to which the need is being felt to develop agroforestry systems, which may be able to meet the requirements of sustainable livelihood in the future. Some of these changes are discussed here.

Heavy Biotic Pressure

Chalcolithic and Iron Age maps of the Indian subcontinent suggest that human occupation of the current Thar Desert region was thin during these earlier ages, and that populations depended upon hunting and limited pastoralism. From the analysis of available census records, the population of the Thar Desert region registered an increase of 490% from 1901 to 1991. The present population of the Thar is about 40 million. Decennial variation in population from 1901 to 1991 in the Thar showed a growth rate of 186% (between 1901 and 1971) as compared with 132% for the whole country. The population growth rate for the decades 1971-1981 and 1981-1991 were 36.7% and 30.7%, respectively. The population density is quite high compared with the global average of 6-8 persons/km^2 for arid zones. The major factor responsible for such phenomenal growth of population in the Thar is a wide gap between birth and death rates. Moreover, social values favor large families and positive sanctions for fertility outnumber negative ones.

Cropping on Marginal Lands

Of the total area of the Thar, only 1.3% is now under forest cover. The net cultivated area increased by 54.1% from 1915 to the early 1980s. This trend in cultivated area increase has further accelerated during the past decade (the 1990s), having increased by 11.7% compared with the amount in the early 1980s. As a consequence of this, barren, cultivable and uncultivable wastelands, permanent pastures, and fallow land have declined substantially (Dhir, 1997). There was a significant decline in grazing land area as marginal lands were brought under cultivation in recent years (the late 1990s). Because of low and erratic precipitation, restricted to a very short period, only one crop, the kharif (monsoon) season crop, is sown each year in most of the Thar.

The common crops are pearl millet, cluster bean, green gram, moth bean and sesame. In the arid zone of peninsular India, in addition to these crops, sorghum, maize pigeon pea and other coarse millets are also grown. It has been observed that the agricultural holdings during late medieval times were only one fifth of those found today. By 1951, much of the potential agricultural land was brought under plow, and the trend since then in already established agricultural tracts has been intensification of farming, and the expansion of agriculture into new lands that may be considered environmentally marginal.

Mechanized field operations (use of tractors) not only increase soil erosion but also reduces the naturally regenerated population of various useful species (Jodha, 1997).

Unsustainable Development Policies

With the advancement of science and technology, efforts have also been made to enhance production in the arid zone. Some of the technologies like sand dune stabilization, wind-breaks and agri-horticulture have produced good results because they are developed in synergism with nature (ecosystem) and natives (people and vegetation).

But some other technologies imported from less environmentally stressed areas have proven to be unsustainable. For example use of tractors adversely affected the natural regeneration of native trees like Prosopis cineraria, and increases wind erosion. Irrigation with waters from the Indira Gandhi canal network has caused problems of water logging and, secondary salinization (Dhir, 1983). Several exotic species either become invasive, like Prosopis juliflora, or didn't prove economically productive, like Faidherbia albida, jojaba etc. The overall impact of these imported technologies, along with high biotic pressure and a fragmented approach to desertification control, can now be observed in terms of loss of biodiversity, increasing desertification, and other adverse impacts. It is now imperative to think broadly and plan holistically for the future of arid zone agroforestry, the basis of sustainable livelihood in the arid zone.

Throughout the arid zones, there is no dearth of problems, but rapidly increasing desertification (some call it land degradation) is a worldwide problem. Land degradation appears to be more of a descriptive term than desertification, as it better describes the process of wind and water erosion, soil salinization, and loss of vegetation cover. It helps in clarifying relationships among climate, topography, soil structure, land use, hydrology, and vegetation. Throughout the arid zones of the world, the major factor responsible for land degradation is soil degradation. At the world scale, the GLASSOD database indicates that in arid zones, more than 60% of soil degradation is caused by wind erosion, with human-induced water erosion accounting for 8% of the phenomenon. At the global level, 349.6 million hectares of land in arid zones are characterized by a light to moderate degree of soil degradation, and 42.9 million hectares have strong to extreme soil degradation. Besides wind and water erosion, over-grazing, high magnitude agricultural degradation activities, over exploitation of vegetation for domestic use, e.g. for fuel, and faulty land use are other factors responsible for ever increasing land degradation in arid zones.

The links between land or soil degradation and vegetation in arid zones are complex. Vegetation and soils display differing resilience to disturbance. Although in arid zones vegetation communities can be readily disturbed and degraded, recovery rates are very slow in comparison with those of other ecoclimatic conditions. Vegetation in arid zones is highty sensitive to natural climatic variability. Decreased production and an ability to lie dormant are two strategies displayed by arid zone plants for survival at times of moisture deficiency. Although an adequate database is not available, two useful relationships between soil degradation and vegetation can be determined. First, vegetation cover may help to indicate the vulnerability of a particular habitat to soil degradation, and second, vegetation cover or community type or plant species may indicate that soil degradation.

POLICY, RESEARCH AND DEVELOPMENT RECOMMENDATIONS

Every ecosystem has some buffering capacity that enables it to remain unchanged for some time if it is overexploited.

While this applies to the arid zone, its fragile nature limits its buffering capacity. Thus, for a few decades the shortcomings of policy were not observed but as the buffering capacity is gradually being eroded, the indicators of unsustainability as a whole, e.g. decreasing natural resources, deforestation and desertification, become visible. For fragile ecosystems, like those that occur in arid zones, one thing needs to be recognized by policy makers and scientists, that is; sustainability of agroforestry in the arid zone depends on the degree of synergism with '3Ns' i.e. Nature, Native vegetation and Native people.

Some of the future directions for arid zone agroforestry need to focus on the evaluation, improvement and encouragement of sustainable traditional agroforestry systems, exploring the possibilities of natural regeneration/rehabilitation of arid zone ecosystems, NWFPs as the basis of sustainable livelihood, rethinking of research priorities, and policy improvement. These will be discussed in turn. Evaluation, improvement and encouragement of sustainable traditional agroforestry systems that have been developed for centuries in harmony with nature. As mentioned earlier, agroforestry systems have been developed and used in this region since time immemorial. This may be because the inhabitants of this region found that no other option was sustainable aside from conserving and harvesting whatever nature is giving. The present forms of traditional agroforestry developed over the centuries through the experiments of earlier generations in all types of climatic variations. The validation of their sustainability is very difficult to establish by conducting short-term experiments lasting only 5-6 years. Thus, there is immense possibility for evaluation, improvement and encouragement of these traditional agroforestry systems.

Exploiting the possibilities of natural regeneration/ rehabilitation of arid zone/arid zone agroforestry as an alternative to failed plantation/rehabilitation strategies. It is a mechanism of restoration or maintenance of ecosystems by nature that trees produce numerous seeds. A small part of this seed population is sufficient to replace the loss of plant population due to various factors. Human beings propagate the plants artificially to compensate for what they harvested form the natural vegetation or to get a higher quantity of the natural products. Those needed in large quantities have already been domesticated, e.g. food crops, fruit trees etc. Still a large number of natural species are wild or depend on natural regeneration. Trees in agroforestry may fall under this category. Natural regeneration as well as artificial tree establishing technologies works comparatively better in humid and irrigated areas. In the arid zone, due to climatological, physiological and sociological reasons artificial establishment has not achieved the required success.

The need for domestication and technology development for sustainable harvesting and value addition for non-wood forest products (NWFPs). Unlike the humid or sub-humid regions where trees are typically grown for timber, the trees are grown in this arid region mainly for non-wood forest products (NWFPs). In the arid zone closed forests are seldom available and agroforestry has been followed since time immemorial in which people conserve and manage whatever seeds of native tree species like Prosopis cineraria and Zizyphus numularia germinate in the crop field or on the common land. After the establishment of trees only annual lopping of branches is done which provides precious fodder during scarcity/drought. Taking the example of P. cineraria, this tree is found in this region at densities ranging from 40 to 120 trees/ha.

This we can say is the natural forest in common or private lands otherwise no dense forest is found as in the high rainfall areas. People lop the branches of this tree in the month of December and it becomes lush green in the month of May-June when no other fodder is available. Simultaneously, pods develop which are used as vegetables. Similarly other tree species provide fodder, pods, fruits etc. The interesting thing is that, during drought when annual crops fail, the yield of NWFPs does not decrease drastically, thus trees provide food security during drought. Besides some trees produce other products like gum, resins, edible/non-edible oils etc. which increase the income of the marginal farmers who are in the majority in this region (Sharma, 2000).

The services provided by the trees in hot arid regions are equally important for moderating the climate. Cultivated in the form of shelterbelts, they reduce wind erosion and conserve soil and water.

In agroforestry systems, trees reduce the solar heat load on crops and also decrease wind speed and soil temperature, conserve soil moisture, increase biological activities in the soil and thereby improve the microenvironment for crops.

The yield of crops grown under P. cineraria is generally higher than sole crops, knowing this fact agroforestry has been followed in this region for centuries. The edapho-climatic conditions in this region are so harsh that every tree improves the environment, it may be at micro level, yet its impact is visible (Sharma, 1998).

The other important NWFPs in these arid areas are plants with medicinal values. Due to climatic specialty there are medicinal plants which are either endemic to this region or their quality deteriorate if they are grow/found in higher rainfall areas.

Research Priorities: Considering the social and ecological demands, policy on agroforestry research also needs to be prioritized. Some of the possible priorities are-In-situ rainwater conservation for tree establishment and for agroforestry more generally.

Water is the most limiting factor for productivity and sustainability in the arid zone. As described earlier, the effect of bringing additional water through canals or ground water has been the cause of unsustainability in the ecosystem. Thus the only possibility remains the conservation of rainwater

(Gupta and Sharma, 1998). Trees are the keystone of arid zone conservation and once trees are established the other conservation processes start automatically e.g. rainwater conservation, development of vegetation cover etc. (Sharma, 2000). Further research is required on:

(i) Development of in-situ rainwater conservation techniques that require less labor and maintenance;

(ii) Development of plant-based water absorbing/retaining materials;

(iii) Initiatives for widespread adoption of rainwater conservation techniques;

(iv) Improvement in traditional water harvesting systems;

(v) Development of simplified windbreak establishment techniques.

Improvement of Policy: There are some important aspects on which improvement of policy is needed for creating a more congenial environment for development. These include:

Increased international funding for arid-zone research and development. The arid regions of the world in general suffer from a vicious cycle of low productivity, a low level of investment and as result, poverty. Greater priority is given at the international level for funding of projects and programs related to timber forestry and rainforest management, perhaps because these activities contribute more to international trade. Very limited funds aimed at combating desertification are available to the arid lands.

Forestry and agroforestry activities in arid regions no doubt contribute less to international trade but they provide substantially support to the livelihood of people in these climatically less favored regions.

There is a need to rationalize these priorities and for the international community to give due importance to the non-tradable benefits of forestry/agroforestry, which are otherwise the basis of sustainability and livelihood.

Suggested areas for increased investment in the arid zone are as follows:

(i) *Organic farming in virgin arid lands*. The philosophy of the traditional systems was to work in synergism with nature and use locally available natural resources. High input agriculture (HIA) on the other hand is only sustainable with assured input supply, e.g. irrigation, fertilizers etc, which is not always possible in drylands. Thus the chances of failure are much greater than in traditional systems during drought. Also due to less use of synthetic chemicals in the past, these lands are comparatively free from the residues of such chemicals. Hence organic farming, based on traditional systems and conservation of local natural resources can be one area for investment without much effort. Organic farming of NWFPs can be the first choice for investment (Sharma, 2001).

(ii) *Eco-tourism*. Investment in eco-tourism can prove to be a large income generator in the arid lands. Increasing numbers of people are spending increasing amounts of time and money enjoying the unique resources of these arid land environments, e.g. spectacular scenic beauty, diverse wildlife population, relatively clean air and other amenities such as large open spaces. Investment in the forestry sector can be in amenity plantations, desert parks, buffer strips, roadside plantations and green belts can be economically as well as environmentally beneficial for the arid zone.

Exploring the potential of arid zone agroforestry as a tool for solving environmental problems.

Conservation of Biodiversity: Biodiversity is the essential safety mechanism of any ecosystem and as the arid zone is considered to be fragile in nature the importance of conservation of biodiversity is further increased. Because of biotic interference and developmental activities, not only is the population of various species reduced but also their natural habitats have been destroyed. Thus the easiest way to save these species is ex-situ conservation with protection. This is possible through arid zone agroforestry, in which all types of plants e.g. trees, shrubs, climbers etc can be accommodated, combining conservation with domestication for increased production.

Desertification Control: Desertification is the result of exploitation of arid zone ecosystems beyond their carrying capacity. The carrying capacity of the arid lands is inherently low thus they are highly prone to desertification. Through soil and water conservation and sustainable management of vegetation, the carrying capacity of natural resources can be enhanced.

Various arid zone agroforestry models, e.g. wind breaks, sand dune stabilization, silvi-pasture, agri-horticulture etc. can be appropriate for soil and water conservation in arid lands (Gupta and Sharma, 1997), and the productivity of lands suffering form desertification can be enhanced and in this way desertification can be controlled. Arid lands can be considered simulated desertified drylands for planning of desertification control.

This can be important for preventive strategies for drylands rather than to control desertification after the process starts. Any technology or strategy that is found suitable to increase the productivity of arid lands, would give better results if applied to the desertification-prone drylands because they have a better environment than the arid lands. Assessing the potential of arid zones as carbon sinks.

In addition to their other benefits, trees on farms or in pastures also store carbon. Arid lands cover almost one forth of the earth's surface and any effort to increase tree populations substantially contributes to their value as carbon sinks. Due to their perennial nature and high root turnover, grasses can also prove to be a good carbon sink.

Silvipasture systems are prevalent in most of the arid zone and when both the tree and grass components are properly managed, an increase in net carbon storage can be achieved. The arid lands are prone to desertification and trees and shrubs can be grown on these lands to prevent further soil degradation and to produce biofuel. These trees also enhance soil organic carbon content and thereby increase soil carbon storage. About 104 million ha land area in the world is affected by strong and extreme soil erosion.

CONCLUSION

Arid zones are known to be fragile ecosystems and interference without knowing the ecological linkages within these systems can result in their degradation, which is often irreversible. Agroforestry as traditionally followed in these regions is a complete, ecologically sustainable livelihood system. To meet both present and future demands, policies need to be supportive of the development of these traditional agroforestry systems based on synergism with nature. There is also a potential to utilize arid zone agroforestry to solve global problems like desertification, global warming (through increased carbon sequestration), and biodiversity conservation, which will require that greater international policy support for arid zone agroforestry.

8

Desertification Control and Rangeland Management in the Thar Desert of India

NTRODUCTION

In India there are about 2.34 million km^2 of hot desert called 'Thar'. It represents one of the most inhospitable arid zones of the world, spreading mostly through western Rajasthan, Gujarat, South-Western Punjab, Haryana and part of Karnataka. About 85% of the great Indian desert lies in India and the rest in Pakistan. About 91% of the desert, i.e. 2.08 million km^2, falls in Rajasthan covering about 61% of the geographi-cal area of the state. The Aravali hills, older than the Himalayas, intersect the State to the north-east and in the west lies the great Indian desert the 'Thar'.

The Indian desert is characterized by high velocity wind, huge shifting and rolling sand dunes; high diurnal variation of temperature; scarce rainfall; intense solar radia-tion and high rate of evaporation. Thar desert receives between 100 to 500 mm of rainfall every year, 90% of which is received between July and September.

The sandy soils of the desert have a rapid infiltration rate of water, poor fertility, low humus con-tent due to rapid oxidation and high salinity. All conditions are very hostile for the ex-istence of life, yet, large human and livestock populations inhabit the area. The Indian desert is highly fragile with poor primary producers but large liabilities i.e. the con-sumers causing severe impediments in its 'ecological regeneration' and 'desertification control' efforts. Archeological evidence suggest that the region was once a flourishing green coun-try-side with thick forests and river-systems of which 'Saraswati' and 'Yamuna' were the most important. Epigraphic evidence by Landsat Satellites Imagery support this evidence. The onslaught of man and his domestic animals on the local ecosystem changed the panorama of the region from a land of plenty to the land of poverty in less than 5000 years. The artifacts discovered in the Pushkar and Luni basin of Rajasthan prove that man lived here as early as 5000 B.C.

Perhaps the over-exploitation of land and water resources since the earliest times has made the Rajasthan desert as a 'man-maintained' if not 'man-made'. The word desert gives the impression of a vast, tree-less undulating expanse of sand. The great Indian desert 'Thar' does not conform to this popular notion. The desert is not an endless stretch of sand dunes, bereft of life or vegetation. During certain periods it blooms with a colourful range of trees and grasses and abounds in amazing variety of birds and animal life. Thar desert is highly 'generic' for it become lush green with slightest precipitation. The soil is full of dor-mant seeds of various species which sprout with little moisture.

After independence, India paid much attention towards desertification control and ecological restoration of the Thar desert. A water canal called the Indira Gandhi Canal was constructed to bring the sweet Himalayan waters to the desert region. The IG Ca-nal, 649 km long, has a capacity to flow 524 m^3 of water/sec. With the canal water, the entire scenario of the Thar desert is changing fast. The 'desert ecosystem' appears to be transforming into an 'ever-green' forest ecosystem in the Command Area of the canal (Sinha 1993).

STRATEGIES TOWARDS DESERTIFICATION CONTROL IN THE THAR DESERT

Desert Afforestation Research Station to Control desertification was set up in Jodhpur in 1952. The State Forest Department made a humble beginning towards desertifica-tion control in 1958 by taking up afforestation in the Command Areas of IG canal in Ganganagar district. The major afforestation programmes implemented were: 'sand dune fixation work'; 'silvipastoral plantations'; 'village fodder and fuelwood plantation'; 'shelterbelt plantation'; 'ecological regeneration; restoration and rehabilitation of degraded desert lands'; 'afforestation on barren hills; re-seeding of old pastures and farm forestry'.

Introduction of Fast Growing 'Exotic' Tree Species

The indigenous tree species growing in the Thar desert are not only few in number but are also extremely slow growing. Therefore, greater attention was focused on the in-troduction and selection of fast growing exotic tree and shrub species from isoclimatic regions of the world. In this effort about 115 Eucalyptus species, 73 Acacia species and 170 miscellaneous ones from various countries including Mexico, USA, Latin America, former USSR, Africa, Israel, Peru, Kenya, Australia, Chile, Sudan and Zim-babwe and the Middle East, were introduced. Acacia tortilis an exotic from Israel for sand dune stabilization, Prosopis juliflora, suitable for fast biomass production, Acacia nubica for sand dune stabilization, Colophospermum mopane and Dichrostachys glomerata for fodder purpose and Eucalyptus cameldulensis are few exotics suited for low rainfall area.

A number of the exotic tree species like Eucalyptus cameldulensis, E. melanophloia, Acacia tortilis, A. cillata, A. raddiana, A. senegal, A. sieberiana, A. aneura, A. salicina, Colophospermum mopane, Dichrostachys glomerata, Brasiletta millis, Schinus molis and Prosopis juliflora have emerged very promising for the In-dian desert. Of all the exotic species tried, 'Acacia tortilis' from Israel has been ad-judged the best 'fuel-cum-fodder species' for the desert. Since its introduction, it has found a 'niche' not only in Rajasthan desert but also in other States of India. Acacia tortilis which has shown performance in growth and survival equal to or better than the indigenous Acacia senegal, is very promising (Muthana and Arora 1973). How-ever the desert dwellers are not happy. They want Acacia senegal to be planted on large scale. This species of Acacia senegal is of great socio-economic value for them. Besides yielding fodder and fuel it also gives a valuable 'gum resin'. They are also somehow linked with the chain of the great Indian majestic bird of the Thar desert called 'godawan' which is threatened with extinction. Godawan feeds on the insects which thrive on the gum resins of Acacia senegal.

STABILIZATION OF SHIFTING SAND-DUNES

In low-rainfall areas (150 mm to 400 mm), huge shifting sand dunes are commonly found, particularly near human habitations. Techniques of afforesting the shifting dunes were standardized after 10 years of experimentation. These techniques consist of: (i) protection against biotic interferences; (ii) treatment of shifting sand-dunes by fixing barriers in parallel stripes or in a 'chess-board' design, using the local shrub material starting from the crest of the dunes to protect the seedlings from burial or ex-posure by the blowing of sand; (iii) afforestation of such treated dunes by direct seed-ling and planting. The two species commonly used for erecting 'brush-wood barriers' (micro-windbreaks) are Zizyphus nummularia and Crotalaria burhia (Kavia and Harsh 1993).

The indigenous and exotic species which have proved successful in sand dune sta-bilization are: trees-Acacia senegal, Prosopis juliflora, Albizzia lebbeck, Cordia rothii, Dalbergia sissoo (in regions with mean annual rainfall of 250 mm) and Zizy-phus jujuba; shrubs-Calligonum polygonoides, Cassia auriculata, Ricinus communis and Zizyphus nummularia; grasses-Lasiurus sindicus, Panicum turgidum and Erian-thus munia. Among exotic species, Eucalyptus oleosa (Australia), Acacia tortilis (Is-rael), Parkinsonia aculeata and Acacia victoriae (Australia), Acacia albida (Middle East) were found to be very promising, especially as these species were found to be frost resistant. Acacia albida was used to stabilize 60,000 hectares of sand-dunes in the Thar desert (Anonymous 1988).

Phog (Calligonum polygonoides) needs special mention. It is a very useful species of the Thar desert. It is a naturally growing shrub on the sand dunes. It has a massive network of underground root which works as effective 'sand binder'.

Other species occurring on sand dunes are Aerva psuedotomentosa, Leptadenia pyrotechnica, Ci-trullus colocynthis, Lasiurus sindicus, Calotropis procera, etc. Other suitable species for planting on sand dunes are Colophospermum mopane and Prosopis cineraria.

SHELTERBELT PLANTATIONS TO REDUCE WIND VELOCITY

Shelterbelts and tree-screens consisting of a row of trees viz. Acacia tortilis, Tamarix articulata and Azadirachta indica flanked by two rows (one on each side) of smaller trees like Acacia senegal, Prosopis juliflora etc., with two rows (one on each side) of shrubs like Aerva tomentosa, Zizyphus spinachristi, Calligonum polygonoides were found to be very effective for Thar desert. In the Bikaner region rows of Eucalyptus cameldulensis and Dalbergia sissoo have also worked as effective shelterbelts and tree screens for creation of 'micro-climates' (Sinha 1993). Shelterbelts reduced the wind velocity by 20-46% on the leeward side for 2H-10H during the monsoon period (H= height of shelterbelt). Soil loss was also considerably reduced. Rajasthan State Forest Department has so far covered about 38,000 row km area under shelterbelt, road side, railway line and canal side plantations since 1978 by adopting the technology developed at CAZRI, Jodhpur (Kaul 1969).

ECOLOGICAL REGENERATION THROUGH AERIAL SEEDING

Aerial seeding of seed pellets of Cenchrus ciliaris, Acacia tortilis and Colophosper-mum mopane was done in 56 ha of military range areas in Barmer district of Rajasthan at two sites: the duny sandy plain of Jalipa and the rocky hill ranges of Jasai village. The programme was undertaken in collaboration with the military. Two methods of seeding, i.e., seeding by helicopter and manual broadcasting of pellets were used. Data on establishment and growth were recorded twice, i.e., at sowing and maturity stages. The manual broadcasting of pellets was found better than aerial seeding by helicopter. Under controlled conditions in the Jasai area, the plant population at sandy and rocky area were 6 and 3 individuals per m^2, respectively. The controlled sites were seeded manually (Anonymous 1995).

Aerial seeding was also done in Bikaner on the left bank of the IG Canal in Sardar-pura (300 ha) and Motigarh (400 ha). The seed mixture consisted of seeds of Acacia tortilis, Colophospermum mopane, Dichrostachys nutans, Prosopis cineraria, Zizy-phus rotandifolia, Citrullus colocynthis and Lasiurus sindicus at the rate of 14 kg/ha. A. tortilis recorded highest germination and seedling density followed by C. colocynthis (Anonymous 1995).

ECOLOGICAL RESTORATION AND REGENERATION OF THE MINED WASTELANDS

Two mined wastelands, one of gypsum and the other of limestone were selected for ecological regeneration. Four plots of one hectare each were demarcated to have four treatments, i.e., control, development of micro-catchment area, half moon structure development and ridge and furrow system, so that the planting pits (60 cm) might re-ceive additional rain water as run-off.

Seven indigenous and exotic species of trees and four of shrubs were selected for plantation. In general, 5 m × 5 m spacing in plant to plant and row to row was adopted. Plant species included Salvadora oleoides, S. persica, Acacia tortilis. Azadirachta indica, Prosopis juliflora, Tamarix articulata, Pithecelobium dulca, Dichrostachys nutan, Cassia stnitii, Cercidium floridun and Caesalpinnea ceraria (Anonymous 1995).

The rooted slips of grasses like Cenchrus ciliaris, C. setigerus and Cymbopogon jwarancusa were also transplanted as a single row in the ridge and furrow system. 1.5 ha of mining muck heaps and the rocky substrata were used for plantation. The muck heaps were reshaped to create slopes and inverted terraces for rainwater harvesting. At the rocky site 'half-moon' structures were developed with 3 m and 5 m spacing. More than 90% of the plants survived. Live hedges and biofences of Prosopis juliflora were raised at both the sites with 1 m × 1 m spacing to protect the plantation.

ECOLOGICAL REGENERATION THROUGH FENCING AND ENCLOSURES

The Thar desert shows tremendous resilience for regeneration when it is protected by fences and enclosures for a certain period of time. A chain of long term enclosure has been established by CAZRI and the State Forest Department on a variety of desert habitats such as hills, rocky gravelly pediments, flat pediments, sandy undulating pediments, flat aggraded older alluvial plains, sand dunes and shallow saline depres-sions (see Tables 1 and 2). The recommended duration of protection is from 5 to 20 years. There is good growth of vegetation, both trees and grasses, upon protection. Desertification is completely halted in such regions.

POTENTIAL FOR RANGELAND DEVELOPMENT IN THE THAR DESERT

Natural regeneration of trees, shrubs and grasses in the Thar desert is very slow. The biotic pressure is so intense in the arid region of Rajasthan that the forest floor is grazed clean and the plants are browsed severely by the huge popu-lation of livestock hampering the regeneration process.

Improved Strains of Pasture Grasses and Legumes for Thar Desert

For achieving a breakthrough in the yield of different promising grasses and legumes, an exhaustive germ-plasm of over 1600 plants has been built up at the Central Arid Zone Research Institute (CAZRI) Jodhpur, Rajasthan and maintained in nursery. An active variety construction programme with manifold objective is under way. A num-ber of promising cultivars have been identified. The important cultivars are: CAZRI-358, 75, Molopo, Buffel of Cenchrus ciliaris; 76, 175, 296 of C. setigerus; 317, M-20-5, 319 of Lasiurus sindicus; 490, 491 of Dichanthium annulatum; 331, 333 of Pani-cum antidotale; 144, 1258, 1462, 1626 of Lablab purpureas; 752, 466, 1433 of Clito-ria ternatea. These varieties provide high and nutritive forage over a greater part of the year. They are drought hardy, stable, persistent and aggressive in rangeland and possess high seed yield ability, fast regeneration and good germination. Several areas of the Thar desert are highly saline which hinders growth of grasses and palatable herbs. Salt tolerance of various pasture species has been identified after yearly monitoring of their occurrence and yield and the results have been successfully utilized in the rangeland development in the salt affected areas of the Thar desert.

Discovery of new high yielding pasture grass varieties in the Thar desert:

1. *Cenchrus ciliaris ('Marwar Anjan'), (CAZRI-75)*: It is a selection from an entry in the germplasm received from Australia. In August 1985 CAZRI released this vari-ety for cultivation in the arid and semi-arid parts of India. Marwar Anjan is a tall, thick stemmed, erect and drought hardy perennial. Its leaves are broad, long, droopy and remain green up to maturity. It has a wide adaptability, high tillering ability and is good in regeneration. It possesses a stout root/rhizome system. It gives 2 to 3 cuttings per year. It yields 70 quintals of green fodder and 30 quintals of dry matter per hectare under desert conditions. It has 8% protein and about 60% digestibility. It produces 1 to 1.5 quintal/ha seed even after one cutting of fodder. It is persistent and aggressive variety for rangelands and remains productive for four to five years under proper management systems. It is sown at the onset of the mon-soon with the seed rate of 5 to 6 kg/ha (Anonymous 1988).

2. *Cenchrus setigerus ('Marwar Dhaman'), (CAZRI-175)*: It is a selection from exotic material and well adopted in the arid and semi-arid regions of India. 'Marwar Dhaman' is excellent for grazing purpose due to its thin stem and leafy foliage. It is a drought hardy perennial grass which forms clumps at the base. It is an early maturing variety, flowers between 45 to 55 days, high in tillering, with good regenera-tion abilities and capable of giving 2 to 3 cuts per year under rainfed conditions. It provides an average yield of 40 quintal/hectare green fodder and 5 quintal/hectare dry matter in desertic

regions, whereas in semi-desertic areas, the yield becomes doubled. It contains 9.5% crude protein and has 65% digestibility at half bloom stage. Its pasture remains productive for 4 to 5 years. It is moderately resistant to major insect pests. (Anonymous 1988).

The use of these improved grasses for reseeding the depleted 'orans' or 'village grazing lands' (gochars) assume significance for meeting the grazing needs of desert livestock.

A TRADITIONAL PRACTICE

Growing trees and grasses together has been a traditional practice in the Thar desert. It is being revived on large scale for rangeland development. Among nine tree species tried, Acacia tortilis showed survival of 98% followed by Dichanthium nutans (88%), Acacia senegal (83%) and Acacia indica (65%), whereas Prosopis cineraria and Al-bizzia lebbeck showed only 5 to 10% survival. Hardwickia binata, Colophospermum mopane and Zizyphus nummularia showed 28 to 35% survival. Maximum height was recorded in the case of A. tortilis (291 cm) followed by A. indica (240 cm) and A. sen-egal (175 cm) and lowest in P. cineraria (77 cm). Collar diameter recorded in A. tor-tilis, A. indica and A. senegal was 8.21, 6.88 and 4.46 cm respectively.

Dry forage yield of Cenchrus setigerus in the interspaces between the trees was 2020 kg/ha in the plots of A. tortilis and 2500 kg/ha in A. indica. Dry forage yield of grass under other tree species ranged from 2750 to 2980 kg/ha as against 3000 kg/ha under pure pasture without trees. Cenchrus ciliaris (CAZRI-75) and Cenchrus seti-gerus (CAZRI-175) were sown between rows of neem (Azadirachta indica), subabool (Leucaena leucocephala) and Israeli babool (Acacia tortilis) planted in 1988 at 5 m × 5 m spacing. The growth of trees were reduced, but growth of grasses were better un-der the system. The fodder yield was more under Azadirachta indica as compared to grass under Leucaena leucocephala and Acacia tortilis (Anonymous 1993).

Acacia leucophloea, A. auriculiformis, A. nilotica, A. senegal, A. tortilis, Albizzia lebbeck, Cassia siamea, Dalbergia sissoo,

Derris indica, Parkinsonia aculeata, Prosopis cineraria and P. chilensis are some important multipurpose nitrogen fixing trees recommended for silvipasture and rangeland development in the Thar desert (Si-vaji et al. 1993). Under the rangeland development scheme, controlled and rotational grazing has been introduced. Some of the areas have been reseeded with perennial and nutritive grasses like Cenchrus ciliaris, Cenchrus setigerus, Dichanthium annulatum and Lasiurus sindicus.

Significant achievement has been made in the mass production of nutritive fodder 'sewan grass' (Lasiurus sindicus) for the desert livestock. This revolutionary grass, besides contributing in the development of good rangeland in the Thar desert, has sig-nificantly helped in stabilizing the blowing sand dunes and expansion of the desert (Sinha 1996). In the Jaisalmer district of Thar desert, schemes have been launched for the cultivation of green fodder in 200 ha, pasture development in 3,000 ha, and the es-tablishment of one hundred 'wood lots' and 'nurseries'.

Village Grazing Lands and Protected Fodder Lands in the Thar Desert

In the earlier days, each and every desert settlement had well guarded common prop-erty land resource such as good pasture land called 'gochar', and natural woodland called 'oran'. The 'orans' (Forests of God) were designated to honour village diety or saint and are preserved meticulously on socio-religious grounds. It is like a mini 'bio-sphere' reserve for the village and the propagules released from gochars and orans.

9

Using Poplars and Willows to Provide Ecosystem Services and Bioenergy

Crown characteristics of forest trees are important traits both for silviculture and also for genetic improvement, but its progress which was made on its improvement was rather small due to the complexity and heavy work for crown traits measurement. Fractal theory provided a new idea and also a method to deal with such complicated traits.

A software (FDC1.0) that was aimed at estimating fractal dimension for any plane picture was developed based on the theory of box-counting dimension, the basic concept and technical approach for development of this software was also described.

The crown characteristics of 28 poplar (Populus deltoides) clones, which were part of tested clones in a clone testing plantation established in Siyang Farm, Jiangsu province in 1997, were investigated in this study. The data were collected twice in 1999 and 2000. The results show that there was no significant differences among clones for these traits that stand for crown shape such as crown width (CW), crown height (CH), crown shape ratio (CSR) and for trait that represents leaf quality as leaf coefficient (LC); however, a significant difference was found among clones for these characteristics that reflects crown size such as crown volume (Vc), the outer surface area of crown (TCA) and leaf area index (LAI).

Pictures were taken for crowns and also for 1st, 2nd, 3rd-order branches by a digit camera and then edited with Photoshop

6.0, Fractal dimensions were obtained for each picture by a software (FDC1.0) which was developed by our group. A similar variation pattern to crown characteristics was found for fractal dimension (FD) of crown, fractal dimension (FD) could be a substitutive criteria for crown characteristics. A stronger correlation was found between the fractal dimension of crown and the fractal dimensions of 1st, 2nd, 3rd-order branches, it brought out that crown of poplar is something of self-similarity.

A STUDY ON EFFECTS OF BIOLOGICAL MULCHING ON GROWTH AND SOIL PROPERTIES OF POPLAR PLANTATION IN SOUTHERN UPLAND AREA

Recent years lots of poplar plantations have been established in the upland area. However, poor soil conditions are the main factors to affect poplar growth. In order to improve the soil properties and increase the productivity of poplar plantations in this area, biological mulching treatments were designed in Guizhou province. Two treatments were arranged in this experiment, e.g. mulching weights (four levels:0.0, 2.5, 5.0 and 7.5kg/m^2 with mixed muching materials) and four mulching materials(Quercus fabri, Imperata cylindrical var. major and Pteridium aquilinum var.latiuscukum,Coriaria nepalensis with muching weights of 5.0kg/m^2). The two-year results are as follows:

There are great differences in decomposition rate and nutrient release rate for various plants. Both decomposition rate and nutrient release rate of Quercus fabric are the largest among the mulching plants. With time passing, decomposition rate of mulching plants and release rate of plant nutrients are slowed down gradually, but its stability varies in different plant species.

Biological mulching in poplar plantation not only affects the soil physical properties, such as soil temperature, soil water content, soil bulk density, but also influences the soil chemical properties (such as the contents of total nitrogen, available nitrogen, available phosphate, available potassium, pH value and organic matter) and nitrogen mineralization process(such as the amount of nitrobacteria, the amount of nitrogen net mineralization). Although some differences exist in changing trends of soil properties, the dynamics of the soil properties could be described by "S" shape curves or polynomial functions after biological mulching of 2-years.

In mulching period of 2-years, the stability in soil temperature, water content, bulk density, total nitrogen, available nitrogen, the amount of nitrogen net mineralization, the amount of nitrobacteria in soil, available phosphate content, available potassium content, pH value and organic matter in soil are changed with the different mulching weights and mulching plants.

Generally, the soil are divided into three layers based on the amount and stability of physical and chemical properties in soil, eg, changeable layer of 0-5?, transforming layer of 5-20?,and steady layer of 20-40?. The stability of physical and chemical property in soil varies according to different layers.

Leaf water content, leaf area, nutrient and chlorophyll contents of the leaf are affected by the biological mulching, furthermore, the tree height, diameter at breast height, root grow and total biomass are also influenced. Compared with the control (ck), poplar growth is significantly improved by biological mulching. However, the growth increments vary among the treatments, for example, compared with the control, the increment of tree height and DBH(diameter at breast height) in the treatment with 7.5kg/m^2 mulching is about 2.21 and 1.94 times respectively,

Based on the result from this study, we suggest that best growth increment are found in the treatment with 7.5kg/m2 mulching for mulching weight, while best growth increment was observed in the treatment with Quercus fabri, Imperata cylindrical var.majoris, and it is suitable for them to select in developing poplar plantation in Guizhou upland areas.

BIOMASS PRODUCTION FROM SELECTED POPLAR CLONES: 3-YEAR-OLD EXPERIMENTAL PLANTING

In this work the results of growth and biomass production of ten poplar clones, 3-year-old, are presented. Clones were established in an experimental planting at the forest nursery of Peponia (prefecture of Serres) in N. Greece, at 1.0 x 1.0 m spacing, using the standard method of hardwood cuttings. Poplar clones used in this study were (hybrids P. deltoides x P. nigra or selected clones of P. deltoides) the following: ?-74/76, ?-77/74, ?-81/74, ?-102/74, BELLOTO, TIEPOLO, CIMA, I-214, I-45/51 and He-X/3. The aim of this field trial was the production of biomass, using the coppice system and rotation period of 3 years.

The results at the end of the first rotation period (1999) demonstrated that diameter growth at breast height (DBH), total height (H), production of fresh biomass (FW) and dry biomass (DW) differed significantly between clones. Phenotypic correlations (Pearson's correlation coefficients) between DB (diameter at ramet base), DBH, H, FW (Stem), FW (Branches+Leaves), and FW (Total) were also calculated for each clone separately, and were found statistically significant (r = 0.70-0.99, P =0.01). DBH growth was found highly correlated with biomass production and clones with the highest diameter growth (e.g. I-81/74, He-X3) achieved the highest biomass yields. Under such a close spacing (1.0 x 1.0 m) the known and widely used clone ?-214 produced the lowest biomass yield, whereas a significant number of other clones achieved acceptable biomass quantities.

DEVELOPMENT OF COMPREHENSIVE EXPERT SYSTEM ON

By applying the stratified module design to the Powerbuild and its data bank system, poplar pest expert consultation system was developed. With the system, 163 poplar pests in Shandong Province could be distinguished and their adult bodies could be plotted based on the adult pest form while more than 90 poplar pests could be identified and their larva bodies and injury to the host be plotted based on the larva body form and the injury to the host. With the system developed, occurrence of the most dangerous poplar pests could be predicted and forecasted based on the current situation of the pests and corresponding strategies could be taken to control further spreading of the pests. In this way, it saves the target user much time from the laborious task, e.g. pest identification, technical consultation, prediction and forecasting, control methods and the selection of pesticides.

EFFECTS OF THE WATER EXTRACT OF POPLAR ROOTS TO SEEDS GERMINATION AND SEEDLING GROWTH OF SEVERAL CROPS

The water extract of the roots of two poplar clones, Nanlin 95 and NL-80351,was used to treat the seeds of wheat, rape, cabbage, and cotton. The results showed that the water extract could affect seeds germination, seedling height and root length of wheat, rape and cabbage, especially at a high concentration. The affected intensity to wheat was strongest, and to rape was in the medial. The water extract promoted the seeds germination, seedling height, and root length of cotton significantly. Regarding to the suppression to seeds germination, seedling height, and root length of wheat and rape, and to the promotion to seeds germination and seedling height of cotton, the water extract of Nanlin 95 clone was stronger than that of NL-80351 clone. But the promotion to root length of cotton was on the contrary. The effects to cabbage were a little different to those of another. As to the suppression to seeds germination and root length, the water extract of Nanlin 95 clone was stronger than that of NL-80351 clone. And to seedling height, contrary conclusion was drawn. The water extract of Nanlin 95 clone suppressed it, while that of NL-80351 clone promoted it. The conclusions can be referred to the species selection of agroforestry systems.

GENETIC CONTAINMENT OF POPLAR PLANTATIONS

Dispersal of pollen, seeds, or vegetative propagules from intensively bred, exotic, or recombinant DNA-modified plantations can in some cases cause significant ecological impacts on wild and managed ecosystems. Insertion of genes designed to prevent or substantially reduce one or more of these dispersal routes can provide major leverage points to reduce the risk and extent of impacts. Containment measures may also be required by law or marketplace constraints, regardless of biological impacts or technology benefits. We will first discuss the rationale and context for choosing genetic containment or mitigation systems, and assessing their adequacy.We will then present data on:

A landscape-based population genetic modeling approach to identifying the level of containment needed and, thus, for determining how well a genetic system must perform to meet environmental objectives. We found sterility genes provided high levels of containment in most simulations, including when sterility was incomplete, unstable, or linked to a gene that provided a selective advantage in the wild.

Stability of transgene expression and RNAi suppression from a field study of >400 transgenic events containing reporter genes. We find that transgene expression is predominantly stable under vegetative propagation in poplar, increasing confidence that a transgenic containment system can be reliable.

A field trial from an attempt to create an ablation system based on the poplar LEAFY gene promoter and the barnase cytotoxin and barstar cytotoxin-attenuation genes. We find that deleterious effects of barnase can be substantial, and that field studies are essential for detecting these and other abnormalities. Careful promoter and vector choice are also critical for system design.

RNAi transgenic poplars with a large number of floral homeotic and floral regulatory genes. We find that because several of the target genes are vegetatively expressed, it is possible to predict their future effects on flowering by performing quantitative RT-PCR on vegetative shoots-important for a commercially efficient system.

NITRATE PHYTOREMEDIATION POTENTIAL OF DIFFERENT POPLAR CLONES

Researches conducted in this work involves with ability of different poplar clones for nitrate phytoremediation, and determination of growth and physiological parameters suitable for use in selection of clones for nitrate phytoremediation. Poplar cuttings were grown hydroponically grown in greenhouse under three different concentrations of nitrates: (1) in modified Hoagland solution with 2 mM NO_3, (2) in standard Hoagland nutrient solution with 10mM of NO_3, and (3) in modified Hoagland solution with 30 mM NO_3. Following clones were selected from spectrum of genus Populus: (1) Populus x euramericana cl. Pannonia, (2) Populus deltoides cl. B-81 and (3) hybrid between P. nigra x maximowitzzi and P. nigra var. Italica cl. 9111/93.

During growth of plants, besides the biomass production of plants and leaf area (LA), physiology of different leaf age groups (LPI) was investigated through measuring parameters related to nitrate assimilation such as nitrate reductase activity (NRA), nitrate accumulation in leaves and roots and total nitrogen content in leaves and roots.

Results showed different reaction of poplar clones to given nitrate concentrations, where clone B-81 (Populus deltiodes) showed best results and stability of investigated parameters under different treatments, what indicates its ability for nitrate phytoremediation.

PHYTOREMEDIATION APPLICATIONS WITH POPLARS AND WILLOWS: A WORLDWIDE OVERVIEW

Phytoremediation (phyto) is an emerging green technology for remediating contaminated sites. Phyto involves the use of plants to efficiently and economically clean up contaminated soils, sediments and groundwater. Poplars and willows are being used for phyto in temperate regions worldwide because of their rapid growth, large root systems, unique biochemical processes, and ability to take up large quantities of water. They have proven effective in remediating heavy metals, organic solvents, petroleum products, herbicides/pesticides, explosives, radionuclides, and ammonia/nitrogen.

Most phyto applications use human installed streamside buffers, vegetative filters, vegetative caps, and in-situ plantings. The poplar/willow clones and/or species must be properly matched to the soil, microclimate, region, and contaminant of the site. A phased phyto strategy is recommended including screening, verification, demonstration, and implementation. This overview cites examples of successful phyto applications with poplars/willows from Bulgaria, Canada, China, Estonia, Korea, New Zealand, Poland, Sweden, Turkey, UK, Ukraine, and USA. Promising new research results also suggest a bright future for phyto with poplars and willows.

Poplar, a Species to restore salt affected and degraded lands? Physiological comparison of adaptative traits among different species under salt stress conditions.

A large part of desert and degraded world's lands are often affected by soil salinity. Sustainable reforestation activities are important to control soil erosion and fertility losses, to restore salt affected and degraded lands and to improve carbon sequestration. For these reasons identification of fast growing cover plant species resistant to drought and salinity is fundamental.

Xerophytes and halophytes species (Tamarix jordanis and Tamarix aphylla) and a salt tolerant clone of the glicophytes Populus alba (14P11), all coming from different semi-desert areas, were collected and studied in controlled environment. Different salt stress experiments were carried out to investigate specific mechanisms of salt resistance and estimate potential carbon sequestration of these species by measurements of gas exchange, fluorescence, biometric parameters, analysis of carbon stable isotopes and osmolyte concentrations. The work allowed the characterization of species in relation to physiological adaptive traits, WUE, productivity, aptitude to restore degraded and salt affected areas in a short-time, suitability for sustainable reforestation and afforestation activities.

POPLAR-BASED AGROFORESTRY IN ARID/SEMI-ARID REGIONS: OPPORTUNI-TIES AND CHALLENGES IN A GROWING WORLD

As population pressures mount, world supply of timber and wood products will increasingly shift from natural forests to plantations in agricultural settings. Short rotation woody crops will play a significant role in this shift and due to its rapid growth rate and productivity, interspecific clones of the genus Populus will be at the forefront.

Multiple opportunities exist in the arid/semi-arid Four Corners region of the southwestern United States:

(1) The Navajo Agricultural Products Industry (NAPI), a commercial farm operated by the Navajo Nation, has sizable holdings of idle rectangular fields that would benefit from conservation interventions;
(2) Western Excelsior Corporation, a nearby fiber mill, is seeking sustainable alternatives to logging native aspen (Populus tremuloides) from the nearby San Juan Forest;
(3) The Public Service of New Mexico operates a nearby coal-fired power plant which could be retrofitted to co-fire woody feedstock; and increased interest in credit systems based on carbon sequestration of woody biomass could generate interest in the Kyoto protocol.

With these opportunities in mind, a research program was established in 2002 to investigate production challenges of drip irrigated hybrid poplar. In 2002 and 2003, 10 and 20 hybrid poplar clones, respectively, were planted to determine establishment, leaf chlorosis, growth, and survival on a calcareous soil under drip irrigation at the NMSU Agricultural Science Center in Farmington, New Mexico. A range of responses to local environmental and edaphic conditions has been observed. In the 2002 trial, the clone OP-367 had the greatest height (11.8 m), DBH (15.8 cm) and wood volume (88.5 m3ha-1) after four growing seasons. In the 2003 trial, clones 059-289, 184-411, 184-401, OP-367, and PC-06 showed the greatest heights, DBH, and wood volume after two growing seasons but varied in degree of leaf chlorosis as measured with a Minolta 502 SPAD meter. All clones responded positively to iron chelate injected through the irrigation system but hybrids from P. nigra crosses (OP-367, DN-34, and 311-93) were the least chlorotic in both trials. P. nigra appears to infer greater Fe uptake efficiency for hybrid poplar grown on high pH soils.

Two additional studies were planted in 2005. The first study was to determine the feasibility of land applied municipal sewage sludge (biosolids) to remediate Fe chlorosis of hybrid poplar. The clone OP-367 was cultivated under composted biosolids (city of Albuquerque, New Mexico) at two rates. After the first season, biosolid-amended plots produced trees that were tallest, had greater basal diameters and remained the least chlorotic compared to the control and the Fe-fertilizer check plots. The second study was a collaborative trial with GreenWood Resources, Inc. to investigate several factors that apply specifically to the use of woody biomass as fuel or feedstock. Five of 66 tested clones had basal diameters greater than OP-367, 31 exceeded it in DBH, and 37 exceeded it in height. It appears that this wider introduction of germplasm may include several clones that are perhaps better adapted to our arid climate and high pH soils.

Based on our evaluations, the clone OP-367 appears to be a good candidate for local wood production and composted biosolids could represent a cost effective alternative to expensive chelated Fe fertilizers currently providing Fe nutrition. Although OP-367 seems adapted to this arid environment, several other clones are emerging that may be superior to this successful hybrid; this will confirmed after testing for five full growing seasons. There appears to be ample opportunities to use poplar for a range of environmental services in agricultural, industrial, and municipal sectors and to develop of a renewable, biomass feedstock industry to meet the world's growing energy needs.

EARLY RESULTS ON NEW HIGH YIELDING POPLARS CLONES FOR SHORT ROTATION FORESTRY AND PHYTORIMEDIATION

Bioenergy could cover a significant percentage (10%) of the total domestic annual request of energy in Italy, reducing imports of fossil fuels, and respecting CO2 emission levels within the subscribed Kyoto Protocol. Part of the biomass (less than a 20%) may be produced by coppice plantations of fast growing trees (poplars, willows, robinia and eucalypts), established on agricultural lands, and intensively managed with short rotation harvesting cycles (1-5 years). Specific public grants of the European Union (Reg. 1257/1999) have encouraged farmers, during the last 5 years, to establish about 5000 ha of Short Rotation Forest (SRF) plantations, mostly in northern Italy, with poplar clones. Farmers are interested in SRF for those grants, and also because they are seeking new crops, alternative to surplus food crops. Improved planting material for SRF plantations is urgently needed in order to improve the profitability of SRF.

Poplar clones are under investigation in a network of experimental plots in northern Italy, comparing new clones, specifically selected for SRF, with the old ones (such as I214) not specific for SRF. According to the encouraging results of the first biannual cycle, the new poplar clones, such as Monviso, AF2, AF6 and 83.148.041, produce almost 100% more than the old clones.

Maximum yield reached more than 20 t ha-1 year-1 in dry matter. Although these encouraging results, profitability of SRF is still strongly dependent of public grants. Research is in progress in order to combine high biomass production and phytorimediation with disposal of animal sewage sludge in poplar SRF plantations. This could dramatically improve SRF biomass yield and decrease its production cost. Amount of sewage disposal will depend, among other factors, on poplar clones (biomass yield and nitrogen use efficiency) and rotation cycle (1 versus 2 years).

POPLAR TRIALS IN SPAIN FOR BIOMASS AS A RENEWABLE

The EU renewable energy policy establishes the necessity to diversify energy sources and reduce CO2 emissions via the substitution of fossil fuels for renewable energy, and sets an objective of at least 12% of total energy consumption to come from renewable sources by 2010. In Spain, the balance of the policy up to 2004 indicates a significant rise in the overall consumption of renewable energy although as yet, this rise is insufficient, having reached only 28.4% of the total targeted increase (IDEA,2004).

The use of biomass to produce energy is an option which, in spite of its high potential and forecast production of around 150 mtep by 2010 in the EU as a whole (EU biomass action plan 2005), is being developed in Spain at a slower rate than that required to meet the specified objectives.

This is basically due to the high degree of heterogeneity regards the materials to be used and their energy producing potential, with growth in this direction being conditioned by the development of a stable supply market (IDAE, 2004).

Among the many potential sources of biomass (forestry, agriculture, industry, etc..), it is considered that energy crops will constitute around 37.86% of the total increase in primary energy for the period from 2005-2010, with this resource assuring a more stable supply. Woody species are considered especially interesting due to the combustion qualities of their biomass.

Of the many forest species which might be used to establish in short rotations coppice to obtain biomass for energy production, Populus spp. is among those with the greatest potential in our country. This crop would provide an alternative, profitable use for surplus agricultural land and therefore favour the development of the rural areas where crops are established.

It is well known that Populus spp. has excellent characteristics for the proposed use, such as its wide genetic base, short breeding improvement cycles, ease of vegetative reproduction, capacity to produce new shoots after pruning, etc. However, to successfully establish plantations of Populus spp. in Spain, much research is required into areas such as the adaptation of clonal material to the different edafoclimatic zones, adaptation of the relevant silviculture to our particular conditions (optimum density, fertilization, weed control, appropriate rotation cycle,...), the total biomass production, harvesting techniques, transportation, management and use of these resources as well as its potential as a fuel.

To achieve this objective, several projects have been launched such as the "PSE-Energy Crops", considered singular and strategic and financed by the Spanish ministry of Science and Education, in which the use of Populus spp., among many other crops, is contemplated in the subproject 2 (Biomass Production), or "The cultivation of Populus spp. in short rotation coppice to obtain biomass for energy", financed by the Nacional Subprogram of Agricultural Resources and Technology of the INIA.

In both projects, 18 experimental plots have been established in 2006 at seven different locations (Castilla-León, Cataluña, Castilla-La Mancha, Extremadura, Andalucía, Navarra and Madrid), each of them within the potential scope for poplar cultivation in Spain. The size of the trial areas varies from 4ha to 0.25ha, according to its purpose, which might be to estimate production on a real scale or to introduce different experimental variables such as clonal response, spacing, fertilization or rotation cycles.

The clones included in the trials belonging to different species or hybrids of the genus are: 'I-214', 'I-MC','2000 Verde', 'Guardi', 'AF2', 'NNDv', ' A4A' and ' Triplo' (P. x euramericana), 'Unal', ' Raspalje' ' Beaupre' and 'USA 49-177' (P. x interamericana), 'Monviso' and 'Pegaso' (P. deltoides x P.trichocarpa) x P.nigra) and ' 111', 'GU1 18-21', 'PO10 10-20', 'J1 3-18', 'Al 29 8-35' y 'J3 1-3' (P. alba). The densities tested in the trials were 15.000, 20.000 and 33.333 cuttings/ha and the cutting cycles will be 1+3 or 3 years.

Growth and production estimates are made taking into account different types of analysis through dendrometric variables, parameters of plant architecture, ecophysiological variables and biomass estimation. By integrating the appropriate variables, the growth and production of the standing biomass may be modelled.

The intention, therefore, is to make progress regards the suitability of the genus to our country as a producer of biomass for energy purposes, whilst at the same time providing advantages for the environment (reduction of greenhouse gases) as well as social advantages (rural employment).

POPLAR WOOD PROPERTIES ARE AFFECTED BY NITROGEN HYPER-FERTILIZATION

In order to sustain the increasing demand for wood products, foresters are turning to fastgrowing species and plantations. Application of fertilizer to forest stands increases growth rates and biomass but generally reduces wood density. Populus species and their hybrids exhibit a marked response to nitrogen availability. However, surprisingly little is known about how nitrogen availability affects wood formation. In this project, we are investigating the effects of nitrogen availability on wood formation and wood properties in poplar. To induce a marked physiological response, we applied exogenous three levels of ammonium nitrate to greenhouse grown Populus trichocarpa x deltoides cv. H11-11. Limiting (0 mM), adequate (1 mM) and luxuriant (10 mM) levels of ammonium nitrate were used and the xylem was analysed after 28 days.

The deposition of lignin appears to be reduced in trees supplied with luxuriant nitrogen levels, based on a histological investigation using lignin staining. Electron microscopy showed that the poplars under high nitrogen fertilization possess an additionnal cellulose layer reminescent of the G-layer of tension wood. It seems that the fibers produced in these trees were shorter and thinner than normal fibers. Differences in chemical composition were observed between wood samples obtained from the different nitrogen treatments were revealed by thioacidolysis. The frequency of the main lignin units varied according to the level of nitrogen., the syringyl units decreasing under the luxuriant treatment. Changes in structural carbohydrates were also measured and, differences in cell wall composition were confirmed by a preliminary FTIR survey. Microarrays enabled the identification of genes associated with wood formation that are diffentially expressed according to nitrogen availability. In response to elevated nitrogen, several genes participating in lignin biosynthesis showed decreased transcript levels, while a number of genes involved in cell wall carbohydrate assembly were increased. The expression profiles of the low and luxuriant treatments were also compared to those from tension wood produced in leaning trees.

Our results demonstrate that poplars receiving a high nitrogen supply show a delayed xylem development and, relative to untreated trees, a distinct partition of carbon resources into wood components. Consequently, we are proposing that nitrogen availability alters the formation of wood and its properties.

POTENTIAL IMPACT OF GENE FLOW FROM POPULUS X CANADENSIS ON NATURAL RESOURCES OF POPULUS NIGRA L.

We evaluated the potential risks associated with the use of cultivated poplar plantations of P. x canadensis (hybrids resulting from a cross between P. deltoides and P. nigra) and P. nigra cv. italica, in the surroundings of P. nigra in Belgium. First, the flower biology of trees of P. x canadensis, P. nigra cv. italica and P. nigra was studied. The results indicated that flowering time did not constitute a barrier to hybridisation between P. x canadensis and P. nigra.

In order to assess the level of inter-specific hybridisation, the open pollinated offspring of black poplar females was studied by conducting a paternity analysis based on the combined data of different molecular markers (isozymes, SSRs, the diagnostic STS marker win3). This was done on two different study sites: (i) a species mixed poplar plantation where the black poplar females were surrounded by P. nigra as well as P. x canadensis males, and (ii) an open agricultural landscape where the black poplar female was surrounded by many P. x canadensis males but isolated from black poplar males.

In the first study site, introgressive hybridisation was not detected. This was in conformity with several former studies investigating the introgression of genes of P. deltoides in the offspring of black poplar females located in natural black poplar populations. In contrast, introgression was clearly detected in the offspring of the isolated black poplar female located in the open agricultural landscape.

This was the first time that evidence for introgressive hybridisation was presented. Based on the contrasting results obtained on both study sites, we have put forward the hypothesis of pollen competition in P. nigra. The hypothesis of pollen competition between P. nigra and P. x canadensis states that, in a mixed pollen cloud, pollen of P. nigra may be more successful than that from P. x canadensis in pollinating female black poplars. Also, if no pollen of the own species is present, P. nigra females are pollinated successfully by pollen of P. x canadensis. This hypothesis was tested in an experimental study. We analysed the genotypes of seeds produced in a series of controlled hand-pollinated crosses in which pollen of the two species were mixed. Paternity analysis based on the SSR markers confirmed the hypothesis; we found that pollen of P. nigra were more successful in fertilizing P. nigra ovules then pollen of P. x canadensis.

This explains why no introgression was detected in the species mixed poplar stand and explains also the results of former studies reporting on very low levels of introgressive hybridisation in natural populations of black poplar.

Following the finding that pollen of P. nigra and P. x canadensis differ in fertilizing ability of P. nigra ovules, we determined whether delays in pollen arrival time on P. nigra stigma's altered the average success of fertilization. Preliminary results of these pollination experiments will be discussed.

Based on the results, recommendations are made to conserve and restore natural resources of black poplar populations.

PRODUCTIVITY POTENTIAL OF NEW CLONES OF POPLARS IN INDIA

Large scale plantations of poplars being undertaken in India to supplement the wood requirements are the culmination of efforts started in 1950 onwards with introduction and testing of poplar clones from England, Netherlands, Italy, and Germany which got impetus after introductions from Grafton, Australia in late seventies. The Australian selections yielded highly promising results in Tarai regions of Uttaranchal, which prompted large scale plantations in foothills, and northern belt of India in states of Punjab, Haryana and Uttaranchal during eighties. These clones were restrictive to about 4 clones (G-3,

G-48, D-121 & S7C16) and were suitable for Shiwaliks and Outer Himalayas which had great potential for their cultivation. Therefore, the efforts to develop indigenous clones were started in the Parmar University, Solan from 1976 onwards with first Himalayan poplar (Populus ciliata), then with Eastern cottonwood (P. deltoides) and subsequently hybrids of these and other species were developed and now with white poplar (P. alba).

Productivity levels of clones of these various species vary according to site conditions and type of rooting ability and rooting behaviour. Clones have been categorized into five categories depending upon the plunging and anchor root behaviour, and thus dividing them into plantation categories.

The productivity levels vary with various rooting categories, clones in question and site conditions from 15-30 m-3 ha-1 yr-1. Alternatively the trees were found to be yielding wood at the rate of anywhere between 0.5 to 1 m3 per tree and the tonnage yield was also 0.5 to 1.0 t per tree. The productivity levels with Australian clones from Punjab and Haryana are reported to be much higher at 25 to 40 m-3 ha-1 yr-1 with some farmers reporting productivity levels of about 60 m-3 ha-1 yr-1 with intensive farming and supplementation of macro and micro nutrients under rigorous silvicultural practices. Whereas, the productive rotation age of Eastern cottonwood ranged between 7-9 years in foothills, at higher elevations in Kullu valley and Nauni (Solan) for P. deltoides and its hybrids it ranged from 9-15 years. The rotation cycle with P. ciliata and its hybrids was even higher at 15-25 years, again dependant on site, soil, moisture level and the clone planted.

RELATIVE SIGNIFICANCE OF INFLUENCE OF TERM OF PREPARATION AND PLANTING ON THE SURVIVAL OF BLACK POPLAR CUTTINGS

Experiments were established at the beginning (second half of March), middle (beginning of April) and at the end (second half of April) of usual period of the establishment of poplar nurseries in this region. Nine genotypes of black poplars were examined in two years on sandy fluvisol. The percentage of cuttings with live shoots The percentage of cuttings with live shoot at the beginning of May (D30), in the second half of May (D40) and at the end of June (D80), as well as the percentage of cuttings that survived till the end of the growing season (SURV) and percentage of cuttings that gave plants higher then 220 cm (SURV220). The influence of genotype, term of cutting preparation and planting, year, their interactions and Residual was evaluated according to their ccontribution to the total variation (based on three-way ANOVA).

Contribution of genotype to the total variation was around 5%, while the contribution of date of cuttings' preparation and planting (mostly around 20%) was the highest after the contribution of Residual. Influence of year was rather weak, and most considerable for SURV220. Significant interactions were genotype x planting date for D30 and SURV, then interaction date x year for D30 and interaction genotype x year for SURV220.

Strong influence of differences among dates of cuttings' preparation and planting suggests the dependence of cuttings survival on characteristics of cuttings' physiological status and soil conditions. Thus, the significance of proper date of cuttings' preparation and planting and application of proper cultural measures in the spring, could contribute more significantly to the cutting survival than the selection of genotypes which root easily. The improvement of rooting of cuttings is an important task in the breeding of black poplars, but the differences among genotypes in their reaction to the conditions of environment should be respected in designing of cultivar-adjusted nursery and plantation establishment technology.

RESISTANCE AND SUSCEPTIBILITY OF NATIVE AND EXOTIC POPLAR SPECIES AND CLONES TO SOME POPLAR PESTS IN IRAN

Iran is classified as countries that have poor forest areas. About 7% of total country surface is covered with forests trees. In order to prepare needs of the country to wood products, poplar plantation has been considered for years as an alternative wood source instead of forest trees in Iran. By this time, more than 150,000 hectares of talented areas are cultured by different native and exotic poplar clones and according to existence of suitable fields for developing poplar plantations, these areas are increasing to a double amount. In addition to the four native poplar species; Populus caspica, P. alba, P. euphratica and P. nigra that exist as wild or planted trees in Iran, exotic clones of the species P. deltoides, P. alba, P. euramericana and P. P. nigra has been imported to Iran in the recent two decades and their adaptability and yield has been studied in different ecological regions of Iran.

Pests and diseases are among limiting factors of polar plantations in Iran; therefore, reaction of native and exotic poplar clones that are planted for studying their adaptability in different ecological regions of Iran-including northern provinces (Guilan and Mazandaran), west and northwestern provinces (East Azarbaijan, West Azarbaijan, Hamedan and Zanjan) and central provinces (Tehran, Isfahan, Markazi and Chahamahal-va-Bakhtiari)-to important poplar pests of these regions was evaluated. In Northern provinces of Iran resistance and susceptibility of native and exotic poplar species and clones to two important pests of the region including Paranthrene tabaniformis and Platymycterus marmoratus were studied.

In Zanjan, Hamedan and Azarbaijan provinces the studies were conducted on Phleomyzus passerinii, Camarotoscena hoberlandti and Melanophila picta, and in central provinces of Iran including Tehran, Markazi and Chahamahal-va-Bakhtiari, the studies were done on Empoasca decedens, Chaitophorus leucomelas, Pterocomma populeum, Phleomyzus passerinii, Monosteira unicostata, Melasoma populi and Melanophila picta. As it is not possible to present all results briefly in the abstract, part of it is mentioned as example and the remaining pests would be discussed in the lecture.

In these studies, all native clones of P. nigra in all ecological regions of Iran are infested and are susceptible to P. passerinii, while exotic clones originated from Turkey were resistant to the pest in all regions. The species P. alba was completely resistant to the pest and different clones of P. deltoides and P. euramericana showed different degrees of susceptibility, from resistant to complete susceptibility. Studying antibiosis and antixenosis mechanisms of different poplar species and clones to this pest supported the results of field studied. Poplar psyllids are generally distributed in Iran and they are considered as important pests of poplar in Iran.

In studies carried out on 13 clones of P. nigra in Zanjan province, the highest density of galls induced by the pest was recorded on the native clone, P. nigra 42/51 with a significant difference with other studied clones. Evaluation of infection rate of different poplar clones of the species P. alba, P. euramericana, P. deltoides and P. nigra to the gall inducing aphids, Pemphigus spirotheca, P. vesicalis and P. borealis showed that all clones of the species P.deltoides, P.euramericana and P.alba are resistant to the pests, while clones of P. nigra show different reactions to the three aphid species. The clones, P. nigra 42/53, P. nigra 47/13 and P. nigra 74/1 are susceptible to all three aphids, while the two exotic clones, P. nigra 62/172 and P. nigra 62/167 were completely resistant.

In natural field conditions the highest infestation rate to Pterocomma populeum was observed on P. nigra 63/135, P. euramericana 561/41, P. deltoides 69/55 and P. e. vernicubensis, while the clones, P. n. betulifoliae, P. e. triplo, P. n. 42/78, P. a. 58/57 and P. e. grandis were not at all infested. The exotic clones, P. euramericana 561/41 and P. nigra 63/135 had the highest infestation in the both studying years and in both seasons, spring and fall, when the aphid is present in the nature. Fall oviposition of aphid was not observed on the clones, P. nigra 42/78 and P. alba 58/57. The highest oviposition rate of the aphid registered on P. d. 69/56, P. e. grandis, P. n. 63/135 and P. e. 561/41 clones and was significantly higher than other studied clones.

Monosteira unicostata that is one of the serious and important pests of poplar in Iran, infested all studied clones of P. deltoides, P. euramericana, P.alba and P. nigra species. The highest infestation rate of the pest registered on P. nigra and P. alba and P. euramericana clones respectively. The two clones, P. alba 44/9 and P. alba 58/57 and especially the exotic clone P. alba. 44/9 were highly infested with the pest. Among the studied American hybrid clones, the highest infestation was observed on P. e. marilandica. Natural infestation rate of different clones of P. deltoides to M. unicostata was less than other poplar clones. Antibiosis and antixenosis resistance mechanisms and tolerance of poplar species and clones to P. passerinii, C. leucomelas and P. populeum aphids were studied in laboratory conditions and their results would be discussed.

ROOT GROWTH OF YOUNG VERONESE POPLARS ON ERODIBLE HILLSLOPES IN THE SOUTHERN NORTH ISLAND, NEW ZEALAND

Poplars are commonly planted on moist, unstable pastoral hill country to prevent or reduce soil erosion, thereby maintaining hillslope integrity and pasture production. Mechanical reinforcement by poplar root systems aids slope stabilisation. Root mass and distribution were determined for three Populus deltoides nigra 'Veronese' trees aged 5, 7 and 9.5 yr planted from 3 m poles at 8 m x 8 m spacing on a hillslope near Palmerston North in the southern North Island.

Most of the structural roots (= 2 mm diameter) were distributed in the top 40 cm of soil. Vertical roots penetrated to little more than 1.0 m, being the depth of the soil above a fragipan. Total structural root mass was 0.57 kg, 7.8 kg and 17.90 kg for the trees aged 5, 7 and 9.5 yr respectively. Structural root length was 79.4 m at age 5 yr and 643.7 m. at age 9.5 yr. Surrounding trees increased root mass density to 3 times and root length density to 4-5 times the contribution of the single tree at 9.5 yr. The study indicated that root development of wide-spaced poplar trees on hillslopes was minimal in the first five years but then increased rapidly. We concluded that the minimum structural root network required for poplar trees to effectively bind soil does not develop until at least five years

SAP STAIN DISEASE OF POPLAR WOOD IN EAST CHINA

Sap-stain is considered a wood disease and a defect in the upper grades of all hardwood species. Sap stain fungi invade and use the contents of poplar wood but do not decay wood; Stain of poplar wood is divided into two types of sap stain and heart stain. Sap stain includes the blue, gray, black, red staining; heart stain only is red stain; both reduce the commercial value of poplar wood. Research in laboratory shows that sap stain caused by Lasiodiplodia theobromae, Dothiorella gregaria, Alternaria spp., Thielaviopsis sp., Cladosporium sp., Chaetomium sp.etc., and red stain of heart wood caused by Fusarium sp. and bacteria.

STUDIES ON CULTURAL TECHNIQUES FOR TIMBERSTAND OF POPULUS TOMENTOSA

Cultural techniques of Populus tomentosa for timber production has been carried out at the plains in west Shandong Province and the results showed that:

(1) With a rapid growth rate, full stem, varieties of P. tomentosa var.'hopeinic','lu50','38','39' were ideal for the establishment of plantations.

(2) Main cultural techniques for Populus tomentosa were: 60~80cm^3 holes in the fully ploughed soil, grade seedlings of 2 years or grade seedlings of 1 year, 1/3 cut of the newly grown stem and total cut of the branches, 20~40cm planting depth, 600m^3.hm-2 irrigation at the beginning of May or June in the normal precipitation year, better effect with yearly 450g nitrogen fertilization and best effect with the mixture of nitrogen and phosphor fertilization.

(3) Intercropping was an effective way to improve the soil condition and enhance the productivity of the forestland. The effect in order was consistent with the input of the intercropped crops, ie watermelon or vegetables>peanut or winter wheat>soybean.

STUDIES ON THE GROWTH AND YIELD MODELS OF POPLAR PLANTATIONS IN SOUTHERN AREA

The main planting area of southern type poplar is mostly mixed clones afforestation. According to the 163 sample plots which came from the pure even-aged poplar plantations in Jiangsu, Shandong, Anhui, Hunan and Hubei, the paper established a growth and yield model system with multifactorial unified description based on the diameter distribution model. Main results of the study are as follows:

1. The growth difference was tested on the main planting clones I-63, I-69 and I-72 of southern type poplar. The results show the growth of diameter, height, volume demonstrated similarities in the 3 clones. The average relative error is less than ±7.0%.
2. Applying sample plots, 7 index classes were divided out (baseline age is 6 years, and the interval of site index class is 2m), and single form and multiple form site index tables of poplar plantations were constructed. Meanwhile, 7 factors, eg. pH value, organic substance content, soil effective depth, soil total-nitrogen content, water table, soil bulk density and soil texture were picked out as predicting factors, and quantification site index prediction table was established so that provides a uniform quality evaluation for non-forested land and forested land.
3. The relationship between basal area at breast height and stand density can be described as Dg=aN-b in un-thinned and with same dominant height plots. After fitting, the basal area at breast height prediction model, eg. Dg=0.9681H01.3614 N-0.032 H00.5402was selected for poplar plantations in southern area.
4. The diameter distribution of poplar plantations in southern area was studied with 5 methods to estimate parameters of Logistic equation and Weibull function. The parameter recovery of Weibull function fits best, the acceptance rate reached 75.0%. Followed by the parameter prediction of Weibull function, the acceptance rate was 72.5%. The parameter prediction of Logistic equation fits worst, the acceptance rate was only 52.5%. Therefore, the author considers that Weibull function was good to fit the diameter distribution of poplar plantations in southern area, and the best method to estimate parameters was parameter recovery.

5. Based on power function Hi=1.3+c0Dic1 and Schumacher height-diameter relationship equation Hi=1.3+c0e-c1/Di, this paper discussed how the parameters (and) vary with different stand conditions. Results show that and in power function were closely correlated with the stand age (A) and the dominant height (H0), and C0 in Schumacher equation is closely correlated with the dominant height (H0). Power function and hyperbola were used to describe the relationship between C0, C1 and A, H0. Eighteen differential difference equations were chosen by different expression's combination. After fitting, the universal height-diameter curve of poplar plantations in southern area Hi=1.3+1.3094H00.9211e-3.7027/Di was constructed.
6. The single timber volume equation V=0.000022 D1.8012H1.3184 was selected based on the fitting of 8 volume equations in common use. Further more, the estimation model of merchantable volume was established for veneer glued wood and fiber wood. It is convenient to apply the yield model system, because the current overall volume of different stand age, site index and stand density combination can be obtained through model prediction system, as long as inputting stand conditions (stand age, stand density and average dominant height etc). Further more, the detailed volume of every diameter grade and different merchantable volumes of wood can be obtained. Adding the dominant height growth model and basal area at breast height prediction model, the future volume can also be forecasted.
7. Based on the interdependence between total biomass estimator and the biomass estimators of stem, branch, leaf etc, the incompatible problem was solved through the structural design of nonlinear model. Volume was combined into the biomass model as variable to solve the incompatible problem between biomass model and volume table.

THE HYDROGRAPHIC SITUATION OF THE DONGTING LAKE BEACH AND THE SELECTION METHODS OF POPLAR AFFORESTATION AREA

Based on the hydrographic data of the Dongting Lake in the several years, turned the relationship of water level and time of flood season there is a mathematics model of the submerge time and elevation. Based on different flood levels, the beaches of the lake were separated into lake model and river course model.

The submerge time and elevation of the lake model and the start or end of the flood the river course model were followed by exponential function y=a ebx. The submerge time and elevation of river course model in other time of flood were followed by linear function y=a+bx. The diameter breast height, height, volume growth of southern clone series of poplar growth on the beach were reduced with the elevation and increased time. The anti-submerge ability will increase with the increase of age. The beach where water flowed less than 30 days a year should plant for high-speed high-yield poplar forest basis. Flowed water in 30 to 65 days a year should plant for industry forest. Flowed water beyond 65 days should not plant poplar unless trenches are dug or planted higher the ridge.

TECHNOLOGICAL INNOVATIONS IN CULTIVATION AND UTILIZATION OF FARM GROWN POPLARS UNDER AGRO-FORESTRY MODELS

Forest cover in India is 67.5 million hectares. The well stocked forests with more than 40% crown density are only 41.7 million hectares. The remaining 25.9 million hectares comprise of degraded open forests.

The mean annual increment of forests in India is as low as 1-cubic meter per hectare per year compared to the world average of 2.1 cubic meter. With just 2.5 % of land area of the planet earth, India has to support nearly 15% of the world's human and equally large and rather un-productive livestock population causing intense biotic pressure on the forest resources. Over the years, supply of industrial round wood and timber from forest areas has been dwindling.

Thanks to India's National Forest Policy 1988, most of the wood based industries like Plywood, Veneer and Paper-pulp are now largely dependent on farm grown Poplar wood. Expansion of agro-forestry plantations and their utilization in wood based industries is critically important for meeting the national needs of timber and non timber forest products and conservation of bio-diversity. Technology based innovative farm forestry plantations with genetically improved, high yielding and fast growing clonal planting stock of poplars, coupled with production, marketing and export of value added products made from poplar wood have tremendous potential for diversification of agriculture, meeting the growing needs of industrial timber on sustainable basis and creating ample employment opportunities for the rural poor. Poplars have already changed the skyline in the North Indian States of Punjab, Haryana, Uttar-Pradesh and Uttranachal. Annually, about 50 million poplars are raised under agro-forestry plantations in these states covering nearly 30000 hectares, yielding 3.6 million cubic meter of wood every year to support the wood based industries in the Country.

My poster presentation is to depict an array of economically viable poplar based agro-forestry and Silvi-mediculture models on farm lands in India coupled with utilization of farm grown poplars in ply, veneer, pulp and hosts of wood based industries. These developments have switched the dependency of wood based Industries from other ecologically important species grown in bio-diversity rich areas to poplars grown on farm lands, ushering in diversification, prosperity and employment in rural areas.

THE EFFECTS OF FLOODING STRESS ON GROWTH AND WOOD MICROSTRUCTURE OF POPLAR IN BEACH LAND

The effects of flooding duration on the growth dynamics of 1~10-year-old individual volume were studied,the results indicated that remarkable variations existed among different flooding conditions,and the variations were increasingly reduced as the age increase. The effects of flooding stress on the ring width,the quantities of fiber cell,the basic density of wood were remarkable,and that on the length and width of fiber cell were not remarkable. The annual ring width of Poplar was positively related to the amount of radical fiber cell,amount of vessel and fiber proportion.

A remarkable restriction on the cambium cell division was formed as the flooding stress strengthened. With the stability of fiber cell size,the reduce of quantities of fiber cell and vessel led to the decrease of radical growth,while the ratio of cell wall to cavity became smaller result from the radial vessel proportion increased,it caused the decrease of wood basic density. 4 growth types were divided from 5 kinds of sites under different flooding duration,and the growth potential of Poplar for 3 growth types adapt to plantation were calculated through growth model.

THE OPTIMAL WAY ON BULK PROPAGATION OF HYBRID OF POPULUS TREMULA×P. TREMULOIDES

The hybrid aspen of Populus tremula ×P. tremuloides is introduced from Finland in 2002, and micro-propagated and planted in the Maoershan Forest Farm of Northeast Forestry University in 2003. The result shows that this hybrid presents good adaptability in the field as exotic species by clonal test of three years. At the same time, how to propagate these materials in productive ways still is a problem for spread and utilization of hybrid aspen in the future, because at present the tissue culture is a main method in hybrid propagation, normally with low yield and high cost, thus blocking its commercial production on a large scale.

In this paper, the hybrid of Populus tremula ×P. tremuloides is used as the material. A series of experiments are carried out to create an optimal technique of the hybrid propagation for commercial purpose, such as establishing optimal tissue culture system, simplifying the medium in vitro rooting and using cuttings instead of tissue culture, and some other step for making the cost lower. The results are as follows:

1. *Rapid propagation by multiplication of axillary buds*: The optimum differentiation medium is NT+BA0.1mg/L+KT0.5mg/L+2%sucrose, in which the rate

differentiation is 100%; a single explant has 6.9 buds averagely; WPM+NAA0.3mg/L+IAA0.05mg/L+2% sucrose is the optimum medium for rooting. In which the rooting rate is more than 93%. The suitable medium for transplanting the plants is humus and sand, which is mixed in proportion of 2:1. The optimum air temperature for transplanting is 18~26?, and the moisture in the air should be kept above 60% by spraying water during the first five days after transplanting. The survival rate of transplants reaches above 90% in the best treatment.

2. *Simplification of propagation technique*: The medium compositions is important factors to affect on material differentiation with the order in water > inositol > sucrose > mass element. By the treatments that cutting is soaked in 20mg/L IBA for 20 min and 100mg/L for 15 min is similar on rooting initiation, the highest rooting rate is 77.3%. The rooting rate can reach 83.86% in sealed plastic bag.
3. *Micro-cutting propagation technique*: The seedlings by tissue culture are used to establish the hedge for collecting the cuttings. The cuttings is sheared in 1cm and treated by plant hormone in concentration 150mg/L of NAA+IBA in 5 min for rooting, which the rooting rate is 93.75%. The best softwood cutting time is May and June.
4. *Estimation of the seedling cost*: The cost per seedling by normal tissue culture is 0.27 yuan. However, the cost by our simplifying technique plus cutting method from the hedge is 0.19 yuan per seedling, which is declined by 29.6%.

THE UTILITY OF POPLAR AND WILLOW FOR RESTORING

Forest restoration activities practiced in riparian zones throughout temperate regions of the globe regularly involve establishment of various Salicaceae species. The accepted use of poplars (Populus spp.) and willows (Salix spp.) for restoration of riparian forests can be attributed to inherent ecological and silvical traits of the family, as well as attributes that add to their commercial value. Poplars and willows can rapidly establish on degraded sites and thrive in riverine environments because of their ease of vegetative reproduction, capacity to root adventitiously, tolerance of soil saturation and short-term inundation, ability to acclimate to sedimentation, and their exceptional growth rates.

Quick establishment of poplar or willow stands on degraded riverine sites serves to stabilize channel banks and soil, trap nutrients, improve soil quality through organic matter inputs, and provide understory environments favorable for establishment of other species, thereby catalyzing restoration of critical ecosystem processes.

The exceptional productivity of poplars and willows allows for marketing of various products including fiber and timber, biofuel and sequestered carbon, all of which can contribute to economic sustainability of forest restoration activities. This presentation will summarize ecological and silvical traits that underscore the utility of poplars and willows for forest restoration in riparian zones, and illustrate several case studies in which poplars or willows were established to address specific restoration objectives in riparian zones of several different regions.

TOLERANCE OF WILLOW (SALIX. SPP) TO CADMIUM IONS IN SOLUTION CULTURE

We studied tolerance of 48 willow clones in 8 willow characters under 5×10-6 Cd2+ treatment in solution culture. The results showed as follows: as for tolerance indexes, bush clone 8-27, 35-16, 2356, 62-1 of 22 were the highest (above 100%) while clone 2333 was the lowest. Among arbor willow clones, clone 2142, 2146 showed higher tolerance but clone194, 795 showed different level tolerance in 8 characters. The tolerance in root number of the willow clones tested showed greater than in other 7 characters. The leaves of some clones were etiolated or speckled.

UNINDENTIFIED PHYTOPHTHORA-NEW THREAT TO HYBRID POPLARS IN CROATIA

- Alluvial plains around rivers which have, after the hydroameliorative measures on the rivers, lost their natural regime of flooding have become suitable stands for poplar plantations.

The longlasting underground water level on this stands is the main factor which has an influence on physical and chemical soil characteristics and soil fertility. In the case of high water levels these stands are flooded and the water stays a long time on the soil surface because natural waterflow back to riverbeds is restricted. All these factors are favorable for Phytophtora dispersal and infection especially in the case of monocultures.

The first individual dead trees have been noticed in the poplar plantations in the region of forests near river Drava in 2001. The region has 400 ha of poplar plantations and it is estimated that 5 % have been affected by dieback. The dieback has spread to the neighbouring trees. The tree dieback is particulary present on sites in microdepresions with long-lasting water on the soil surface. The dieback is constantly spreading to other localities which are connected with water chanells from which high water is drained into microdepressions on the poplar plantations.

First symptoms of tree dieback were crown damage, chlorosis and leaf shedding. The gum exudation from cracked bark is distinctive in spring and summer. Brown lesions are visible under bark. The disease is spreading in concentric circles to the neighbouring trees and the parts of poplar plantations that are in microdepressions are the most affected.

The soil samples with root and root hairs from the base of the affected trees have been taken for isolation of pathogen in laboratory. The isolates have been sent for ITS fingerprinting method. Results of ITS indicate that sequences of this Phytophthora don't match with any of up to this time noted species from gene bank.

BACKGROUND INFORMATION

Himachal Pradesh, called "Dev Bhumi" the abode of Gods & Goddesses, nestles between "Shivalik Hills" in the South and Great Himalayan ranges in the North.The State is bordered by Jammu & Kashmir on North, Punjab on West and South-West, Haryana on South, Uttar Pradesh on South-East and China on the East.

Himachal Pradesh as political entity came into existence on 15th April 1948. It was constituted by integrating the princely States of Chamba, Mandi, Suket, Sirmaur and 26 smaller states known as the Punjab Hill States. Subsequently, in 1954 Bilaspur, which was then a part "C" State, was also merged with Himachal Pradesh. In 1960 District Kinnaur was carved out of District Mahsu which was earlier constituted by merging 27 hill states.

After the re-organization of Punjab in 1966, Shimla, Kangra, Kullu, Lahaul-Spiti Districts, Una Tehsil of Hoshiarpur District and Nalagarh Tehsil of Ambala District were also merged. In 1972 Districts of Kangra and Mahasu were reorganized. District Kangra was trifurcated and Una and Hamirpur Tehsils were made into separate Districts. Some areas of Mahasu District were merged with Shimla District while the others formed the new District of Solan.

The forests of Himachal Pradesh are rich in vascular flora, which forms the conspicuous vegetation cover. Out of total 45,000 species of plants found in the country as many as 3,295 species (7.32%) are reported in the State.

More than 95% of species are endemic to Himachal and characteristic of Western Himalayan flora, while about 5% (150 species) are exotic introduced over the last 150 years.

PHYSICAL AND GEOGRAPHICAL STRUCTURE

Territory of Himachal Pradesh stretches from 30 22' 40" N to 33 12' 40" N Latitude and from 75 45' 55" E to 79 04' 20" E Longitude. The State is bordered by Jammu & Kashmir on North, Punjab on West and South-West, Haryana on South, Uttar Pradesh on South-East and China on the East.

Land Area

Land area of Himachal Pradesh state is 5567300 hect.. The land use pattern in the state is given in Table

No.	Contents	Area (hect.)
1	Total Geographical Area	
	i) By Professional Survey	55,67,300
	ii) By Village Papers	42,67,196
2	Forests	10,86,548
3	Barran and Unculturable Lands	7,75,896
4	Land Put to Non-Agricultural Use	2,24,791
5	Permanent Pastures and Other Grazing Lands	13,40,222
6	Land under Misc. Trees Crops not Included in Area Sown	97,527
7	Cultivable Waste Lands	1,03,942
8	Other Fallow Lands	22,706
9	Current Fallows	55,410
10	Net Area Sown	5,60,154
11	Area Sown More than Once	4,26,184
12	Total Cropped Area	9,86,338

The Forests

The Forests of Himachal Pradesh known for their grandeur and majesty are like a green pearl in the Himalayan crown. This life supporting systems are presently under great stress due to impact of modern civilization, economic development and growth in human and cattle population. According to national Forest Policy, 1988, atleast two third i. e 66% of the geographical area should be under forest in the hilly states like Himachal Pradesh. However, keeping in view that about 20 % of the area is inaccessible and beyond the tree limit, the State Government aims to bring 50% of the geographical area under forest coverhe forests of the State have been classified on an ecological basis as laid down by Champion and Seth, and can be broadly classified into Coniferous Forests and broad-leaved

Forests. Distribution of various species follows fairly regular altitudinal stratification. The vegetation varies from Dry Scrub Forests at lower altitudes to Alpine Pastures at higher altitudes. In between these two extremes, distinct vegetational zones of Mixed Deciduous Forests, Bamboo, Chil, Oaks, Deodar, Kail, Fir and Spruce, are found.

The richness and diversity of our flora can be gauged from the fact that, out of total 45,000 species found in the country as many as 3,295 species (7.32%) are reported in the State. More than 95% of the species are endemic to Himachal Pradesh and characteristic of Western Himalayan flora, while about 5% (150 species) are exotic, introduced over the last 150 years.

ITEMS	UNITS IN SQ. KM
Reserved Forest Area	1896
Protected Forest Area	33004
Unclassed Forest Area	930
Other Forest Area	405
Forest Not Under Control Of Forest Deptt.	751

The State has 32 numbers of wild life sanctuaries.

ADMINISTRATION

The state having in a legislative assembly of 68 members. The State is divided in to 12 districts. In addition there are 71 Tehsils, 72 Blocks, 28 Nagar Panchayats, 2922 Gram Panchayats(December-1997), 48 Towns and 19388 Villages

DEMOGRAPHIC CONTEXT

Population

The total population of the state is 5170877 person (1991-Cencus) with .45 million living in Urban area and 4.72 million person are living in Rural area. The State has sex ratio of 976 female per thousand male. The density of population is about 93 person per sq. Km.

- LANGUAGE: The official language of the state Hindi & Local Dialects.
- BIRTH RATE: The state has the combined birth rate of 23 (Per 1000)
- DEATH RATE: The state has combined death rate of 8 (Per 1000).
- LITERACY RATE: The total literacy rate of the state is 63.88 % with male literacy rate of 75.36 % and female literacy rate of 52.13 %.
- PER CAPITA INCOME: The per capita income of the state is Rs. 9603/-at 1996-97 base.
- HEALTH AND FAMILY WELFARE: The State has 80 numbers of hospitals (1997-98)

ECONOMIC SITUATION IN HIMACHAL PRADESH

Agricultural

The economy of Himachal Pradesh is dependent upon agriculture and its allied activities and any fluctuation in agricultural production affects the growth rate considerably. The over all economic growth in total state domestic product during 2002-03 (Q) was 4.5 percent as the total state domestic product at constant prices (1993-94) increased to Rs.8454 crore from Rs.8089 crore in 2001-02. The state has made significant progress in the development of Horticulture. The topographical variations and altitude differences coupled with fertile, deep and well drained soils favour the cultivation of temperate to sub tropical fruits. The region is also suitable for cultivation of ancillary horticultural produce like flowers, mushroom, honey and hops.

Tourism

High priority has been accorded to tourism industry, which has emerged as a major sector in the development of economy of the state. The Govt. has developed appropriate infrastructure for the growth of tourism involving activities requiring heavy investment and pioneering commercial projects in new areas where private sector may be reluctant to undertake such activities initially.

Power Generation

Himachal Pradesh state Electricity Board is engaged in the investigation and execution of various hydroelectric projects and transmission and distribution of electricity. Himachal Pradesh has launch a reform programme in the power sector. During the financial year 2003-04, 1357 million units of electricity was generated by March, 2004. Due to limited resources available with state Govt., it has approved the participation of private sector in the generation and supply of power and taken up various hydroelectric projects in private sector for implementation.

Information Technology

Information Technology has a great scope for employment generation and revenue earnings. Himachal Pradesh Govt. in this context has prepared an I.T. Vision 2010 with the assistance of NASSCOM. Govt. has proposed a target of exporting I.T. Software and allied services worth Rs.2000 crore unto the year 2005. Commensurate with the IT policy of the state government certain initiatives have already been taken in the state to further growth of IT.

Road Transport Sector

After the formation of Himachal on 15th July, 1948 the network of roads had received top-most priority of the Govt. as there were very limited roads in the State at that time. At present the road network is widely spread in Himachal Pradesh. The increase in road network is fairly a strong indicator of economic development of the Himachal Pradesh in the field of education, health, agriculture, horticulture and the related fields. On the formation of present Corporation on 02.10.1974, the total routes operated by HRTC were 379 which have grown to 1733 routes in 2000-2001 and the fleet strength have grown from 733 to 1728 in 2000-2001. As the road length increased and basic infrastructure for development also kept pace with it, this had direct relation on the growth of the HRTC.

Bus remains the sole mode of passenger transportation as railways have a very marginal contribution in the State. The railway network in Himachal Pradesh has remained static since 1947, except for a small stretch of about 18 Kms. from Nangal to Una, which has been connected by a broad gauge railway line very recently. This is the only expansion of railway network in Himachal for the last 50 years. The narrow gauge lines connecting Pathankot with Joginder Nagar and Kalka with Shimla are so slow moving that a very small percentage of traffic is carried by them at present; thereby leaving the onus of carrying the passenger traffic on to bus transport.

Besides its operation in the entire Himachal Pradesh including tribal districts of the State, HRTC operates its buses in neighboring States of Punjab, Haryana, Rajasthan, Uttar Pradesh, Jammu & Kashmir, Union territories of Chandigarh and Delhi. Further, it has acquired distinction to ply its buses to the highest village of the Asia and also its buses cross through the three Worlds highest passes namely; Bara-Lacha, Kunjam and Rohtang. HRTC is plying its buses in remotest area of the Pradesh, where private operators hesitate to ply the buses.

FISHERIES AND AQUACULTURE

Himachal Pradesh is blessed with vast and variegated fishery resources in the shape of networks of rivers, streams, tributaries, sprawling reservoirs, natural lakes and ponds etc. Mainly classified into riverine, lacustrine, recreational and pond fisheries, the state waters offer considerable potential for the promotion of fisheries and generating employment. About 12500 fishermen families in the Pradesh depend directly or indirectly on this occupation for their livelihood. During 2003-04, cumulative fish production was of the level of 6465 tonnes valued at Rs. 2977.00 lakh from the state's various reservoirs. The total fish seed production from Carp farms of the sate during the year reached 17.52 million. To make fish feed readily available in the state one feed mill was set up at Jachh, district Kangra. The country's first Mahseer fish seed production farm is also being set up at Sidhpur in Mandi district. The requisite stocking of the fish is being done in the newly created Chamera Dam. An amount of Rs. 292.00 lakh have been sanctioned out of which an amount of Rs. 131.00 lakhs has been released by BBMB, NHPC and SJVN or the following 3 Scheme.

(i) Development of Fisheries in Suketi stream of Mandi district

(ii) Development of Fisheries in Chamera reservoir.

(iii) Development of Fisheries in Sutlej river.

ANIMAL HUSBANDRY AND DAIRYING

Raising up of livestock is an integral component of rural economy. In Himachal there is a dynamic relationship between common property resources (CRPs such as forests, water and grazing land), livestock and crops. Livestock depend to a certain extent on fodder and grass grown on CPRs as well as on crops and residues. At the same time the animals return fodder, grass and crop residues to the CPRs and fields in the form of manure and provide much needed draught power. Livestock thus is an important integral to the sustainability of economy of Himachal Pradesh. The contribution of major livestock products during the year 2002-03 was 7.76 lakh tones of milk, 1610 tonnes of wool, 87.00 million eggs and 3487 tonnes of meat which will likely to be of the order of 7.96 lakh tones of milk, 1680 tonnes of wool, 91.00 million eggs and 3600 tonnes of meat during 2003-04.

Milk production and Per Capita Availability

Year	Milk Production ('000tonnes)	Per Capita Availability (grm./Day)
1998-99	724.0	329
1999-2000	740.0	330
2000-01	760.0	343
2001-02	763.0	344
2002-03	776.0	350
2003-04 (likely)	796.0	359

Animal Husbandry plays an important role to boost the rural economy and as such for livestock development programme attention is paid in the state by way of (i) Animal Health & Disease control (ii) Cattle Development(iii) Sheep Breeding & Development of Wool (iv) Poultry Development(v) Feed & Fodder Development (vi) Veterinary Education and (viii) Livestock Census

Milk Based Industries

H.P MILKFED is operating dairy development activities in the state by providing a remunerative outlet for the surplus milk to the rural milk producers residing in remote and far-flung areas. Majority of these milk producers are small and marginal farmers.

The estimated production of various milk products in the organized sector (MILKFED) including milk sold in the market, paneer, butter, ghee and SFM bottles is shown in table

Production of Milk Based Industry

Product	Unit	2000-01	2001-02	2002-03	2003-04
Milk Sold	Lakh lts.	75.00	74.02	77.97	56.88
Paneer	M.T.	24.46	30.60	32.23	42.24
Butter	M.T.	4.02	5.11	5.10	5.49
Ghee	M.T				

10

Cedrus—the True Cedars

GEOGRAPHICAL DISTRIBUTION

The true cedars consist of 4, or fewer, closely related species of tall, oleoresin-rich, monoecious, conifer-ous, evergreen trees, with geographically separated distributions (Maheshwari and Biswas 1970; Liberty Hyde Bailey Hortorium 1976; Arbez et al. 1978; Farjon 1990; Hillier 1991; Vidakovic 1991; Bariteau and Ferrandes 1992; Tewari 1994). The cedars are restricted to the montane or high montane zones of mountains, situated roughly between 15°W and 80°E and 30°to 40°N (Farjon 1990). This discon-tinuous range is composed of 3 widely separated re-gions in North Africa and Asia:

(1) The Atlas Mountains of North Africa, in northern Morocco and northern Algeria;

(2) Turkey, the mountains on Cyprus and along the eastern border of the Mediter-ranean Sea in Syria and Lebanon; and

(3) The Hindu Kush, Karakoram and Indian Himalayas (Farjon 1990).

The 4 species of Cedrus, C. atlantica (Endl.) Manetti ex Carriere, C. brevifolia (Hook. f.) Henry, C. deodara (Roxb.) G. Don in Loud., and C. libani A. Rich., are so closely related that habitual characteris-tics help differentiate the species (Farjon 1990).

Isozyme analysis of Cedrus diploid tissue raises questions about the separation of C. atlantica and C. libani into 2 distinct species, because no distinguishing gene marker was detected (Panetsos et al. 1992). Using allozyme differentiation, Scaltsoyiannes (1999) determined that these 2 species should be separated into distinct taxa. There is disagreement as to the exact taxonomic status of the various cedars, with some authors suggesting that they be reduced to only 2 species, C. deodara and C. libani. This review examines all 4 species.

The cedars are valuable timber trees and are quite striking as specimen plantings in the landscape. The wood of C. libani is fragrant, durable, and highly decay resistant. On a historical note, the ancient Egyptians employed cedar sawdust (cedar resin) in mummification (Maheshwari and Biswas 1970; Demetci 1986; Chaney 1993). Upon distillation of Cedrus wood, an aromatic oil is obtained that is used for a variety of purposes from scenting soap to medicinal practices (Maheshwari and Biswas 1970; Adams 1991; Chalchat et al. 1994; Tewari 1994). The 4 species of Cedrus, 3 of which are planted to some extent in the United States, are C. atlantica (Atlas cedar), C. brevifolia (Cyprian cedar), C. deodara (Deodar cedar), and C. libani (Cedar of Lebanon).

TAXONOMIC CHARACTERISTICS

Cedrus atlantica is a large tree attaining a height at maturity of 9 to 40 m (29.5 to 131 ft), with rapid growth when young, and is closely related to C. libani. The Atlas cedar is distinguished by a taller crown, less densely arranged branchlets, bluish green leaves (needles) that vary from light green to silvery blue, smaller cones 5 to 8 cm in length by 3 to 5 cm in width (2 to 3 in. by 1 to 2 in.), and smaller seeds (Dirr 1990; Farjon 1990; Loureiro 1990, 1994; Hillier 1991). Young trees appear stiff, with an erect leader and an overall pyramidal shape, but with maturity this species assumes a flat-topped habit with horizontally spreading branches (Dirr 1990). Cedrus atlantica is hardy in U.S. zones 6 to 9, with several beautiful cultivars that differ in color and characteristic habit (Dirr 1990; Hillier 1991; Vidakovic 1991). Of special note is 'Glauca'(f. glauca), a spectacular tree with intense blue to silvery blue leaves, used for specimen plantings (Dirr 1990; Hillier 1991).

Cedrus brevifolia is a rare species with slow growth, but eventually develops into a medium-sized tree, attaining a height at maturity of 8 to 24 m (26 to 79 ft). The Cyprian cedar is distinguished from C. libani only by the habitual form, shorter leaves 0.5 to 1.6 cm in length (0.2 to 0.6 in.), and on older specimens the crown is broad and umbrella-shaped (Farjon 1990; Hillier 1991; Vidakovic 1991). Cedrus deodara is an excellent specimen tree. The habit of the Deodar cedar is broadly pyramidal when young, with gracefully pendulous branches (Dirr 1990; Tewari 1994). It can attain a height at maturity of 15 to 50 m (49 to 164 ft).

It is distinguished from the other species by its drooping leader and longer leaves that are 2 to 6 cm in length (0.8 to 2.4 in.) (Hillier 1991). Multi-stemmed crowns occasionally evolve from the higher branches turned erect, but the crown seldom becomes flat-topped, remaining conical or pyramidal (Farjon 1990). Cedrus deodara is hardy in U.S. zones 7 to 8, but young trees in these zones are prone to injury from frosts and cold wind (Dirr 1990). There are many cultivars of Deodar cedar, but 2 outstanding examples are 'Kashmir'and 'Shalimar'. The cultivar 'Kashmir'is a hardy form, tolerating cold winters to-30°C (-22°F), and is characterized by its silvery blue-green foliage (Dirr 1990; Vidakovic 1991). The cultivar 'Shalimar' displays good blue-green leaf color and is the hardiest cultivar planted in the United States (Koller 1982; Dirr 1990).

Cedrus libani is a majestic tree 15 to 40 m (49 to 131 ft) in height at maturity with innumerable scrip-tural and historical references. Cedar of Lebanon has a thick, massive trunk and wide-spreading branches and is pyramidal when young, but develops a flat-topped crown and horizontally tiered branches when mature (Dirr 1990; Farjon 1990; Hillier 1991; Chaney 1993). The dark green foliage, stiff habit, and rigidly upright cones that are 8 to 12 cm in length by 3 to 6 cm in width (3 to 4.7 in. by 1 to 2.4 in.) give this tree its splendor for landscape specimen planting. The morphological differences between C. libani and C. atlantica are small and not entirely constant (Maheshwari and Biswas 1970; Farjon 1990). Cedar of Lebanon is hardy in U.S. zones 5 to 7 (Dirr 1990; Dirr et al. 1993). A geo-graphical form, spp. stenocoma (Schwarz) Davis, dif-fers from the typical Lebanon cedar in having a broadly columnar habit, needle and cone character-istics intermediate between C. atlanticaand C. libani, and being more cold-hardy (Hillier 1991; Vidakovic 1991). There are also several dwarf cultivars of C. libani of interest for use in the landscape (Hillier 1991; Vidakovic 1991).

The male flowers of Cedrus are erect catkins, up to 5 cm (2 in.) in length, whereas the female flowers are erect, cone-like inflorescences, 1 to 1.5 cm (0.4 to 0.6 in.) long, surrounded by needles at the base (Vidakovic 1991). Male and female strobili of the true cedars are typically borne on the same tree but on separate branches (Maheshwari and Biswas 1970; Rudolf 1974; Farjon 1990).

The male cones grow solitary, more or less erect from the short shoots, and bear abundant yellow pollen (Maheshwari and Biswas 1970; Farjon 1990). Depending upon the al-titude, locality, and weather, the pollen is shed late in the year (autumn), relating to the late development of the female strobilus (Maheshwari and Biswas 1970; Farjon 1990).

The female cones are borne sin-gly at the tips of the dwarf shoots, stand erect, and are less abundant than the male cones (Maheshwari and Biswas 1970; Farjon 1990). Although pollination takes place in the fall, the cones do not mature until the second year, requiring about 17 to 18 months for full development (Maheshwari and Biswas 1970; Rudolf 1974; Farjon 1990).

The mature, barrel-shaped cones are resinous and characterized by numerous closely appressed, very broad scales, each containing 2 seeds (Rudolf 1974). The scales are attached to the persistent rachis with a narrowed, petiolate base and dismember from it by abscission at maturity, as in Abies (Rudolf 1974; Farjon 1990). The irregularly triangular mature seed is rather soft and oily, with resin vesicles present on each side of the seed, and has a membranous, broad wing several times larger than the seed (Rudolf 1974; Farjon 1990). Commercial seed bearing of C. deodara begins from 30 to 45 years of age, and good seed crops are usually borne every 3 years, with light crops in the intervening years (Maheshwari and Biswas 1970; Rudolf 1974; Toth 1979; Doty 1982; Tewari 1994).

PROPAGATION

Cones should be collected directly from the trees, be-fore the cones turn brown, or cone-bearing twigs may be cut from standing or felled trees just before ripening is complete (Rudolf 1974; Dirr and Heuser 1987; Singh et al. 1992). A cubic meter (28.4 bushels) of cones weighs from 12.2 to 15.9 kg (27 to 35 lb) and yields about 1.4 kg (3 lb) of cleaned seed (Rudolf 1974).

Cones should be allowed to dry until the scales loosen and the seeds can be removed (Toth 1980a; Macdonald 1986; Dirr and Heuser 1987). It is important to avoid any more drying than is absolutely necessary, because the seeds may be killed. Cones of Cedrus may be soaked in warm water for 48 h to encourage them to disintegrate (Rudolf 1974; Macdonald 1986). Freezing moist cones (as a last resort) will also force the scales to open up (Macdonald 1986). After the cone scales are dry, they can be placed in a cone shaker to remove the seeds (Rudolf 1974), and seeds separated from the de-bris by fanning or sieving (Macdonald 1986).

Dewinging is done by simply rubbing the seeds in a dry cloth (Macdonald 1986). Resin from the resin pockets in the wings can make the procedure difficult if bare hands are used (Macdonald 1986). Purity of commer-cially cleaned seed has ranged from 85% to 90%. The seeds of Cedrusare oily and do not keep well under ordinary storage conditions (Rudolf 1974; Allen 1995). If Cedrus seeds are dried below a critical level, they will not imbibe water in a way that will allow the food reserves to be used by the embryo (Macdonald 1986). Cedrus seed has retained viability for 3 to 6 years when dried to a moisture content of less than 10%, placed in sealed containers, and held at temperatures of-5°C to -1°C (23°F to 30.2°F) (Rudolf 1974; Erkuloglu 1995).

Cedrus seeds exhibit little or no dormancy and will germinate without pretreatment. However, variable degrees of dormancy may be observed within a single lot of seeds (Dirr and Heuser 1987). Seed should be stratified at 3°C to 5°C (37.4°F to 41°F) for 2 weeks (6.5 weeks for C. brevifolia) to give more uniform ger-mination (Rudolf 1974; Allen 1995).

Thapliyal and Gupta (1980) found that 9°C (48.2°F) was a better temperature for stratification than 3°C. Cedrus deodaraand C. libaniseeds are prone to damping-off disease caused by Fusarium,Rhizoctonia, and Pythium species. Therefore, an appropriate fungicide should be used (Mittal 1983; Tewari 1994). Gordon et al. (1991) and Suszka et al. (1996) provide extensive informa-tion on the sampling, purity, viability and germination testing, seedling evaluation, and storage of forest tree and shrub seeds. Specific procedures are presented for a number of species.

The Association of Official Seed Analysts rules for Cedrus (Rudolf 1974) specify germination tests of stratified seed on top of blotters for 3 weeks at 20°C (68°F) (see also Toth 1980a). International Seed Test-ing Association rules, however, specify diurnally alter-nating temperatures of 20°C at night and 30°C (86°F) during the day for a period of 4 weeks (Rudolf 1974).

Light apparently is not required (Rudolf 1974). Tests may also be made in sand flats (Rudolf 1974). Deodar seed stratified at 4°C (39.2°F) in moist sand for 30 days germinated 45% versus 11% without stratifica-tion (Dirr and Heuser 1987). Thapliyal and Gupta (1980) also found the percentage of germination without stratification to vary from 16% to 69%. Singh et al. (1992) found that seeds from larger-sized cones exhibited higher seed germination (66%) in Hima-layan cedar. Singh et al. (1997) also found significant differences between tree diameter classes in fresh and dry weight of seeds, and germination in the laboratory and in the nursery. Germination of Cedrus seed is epigeal.

Cedrus deodaraseed should be sown in the fall or spring at a rate of 200 to 250 seeds per square meter (10.8 ft^2), in drills 10 to 15 cm (4 to 6 in.) apart, pro-ducing about 19 to 23 seedlings per 0.1 square meter (1 ft^2) for lining-out stock and for rootstocks (Rudolf 1974; Macdonald 1986). Chandra and

Ram (1980) recommend sowing Deodar seed at a depth of 1 cm (0.4 in.). Further increase in depth results in decreased germination. Al-Ashoo and Al-Khaffaf (1997) reported that the best treatment for germination of C.libani seed was a 1.5-cm (0.6-in.) sowing depth, with a covering medium of clay or alluvial soil. In northern areas, fall-sown beds should be mulched over winter, with the mulch removed early in the spring and the bed racks covered with burlap on critical spring nights to prevent freezing (Heit 1968). Cedrus seed can be sown in containers in the fall, transplanted into other con-tainers during the winter, and kept in shaded beds in the summer to produce stock (Rudolf 1974). The size of the propagation con-tainer, growth media, transplanting date, and handling of seedlings is important in container-or field-grown stock (Toth 1980b; Doty 1982; Appleton and Whitcomb 1983; Guehl et al. 1989; Puxeddu and Alias 1991; Burger et al. 1992).

Cedrus deodara 'Shalimar'can be propagated by cuttings (67% rooted) collected in late fall to early winter, when given a quick dip in 5 g/L (5,000 ppm) indole-3-butyric acid (IBA) solution and placed in a sand:perlite medium with bottom heat maintained at 24°C(75°F) (Nicholson 1984). Shamet and Bhardwaj (1995) reported 69% rooting of C. deodara cuttings treated with 5 g/L indole-3-acetic acid-talc or 10 g/L (10,000 ppm) naphthaleneacetic acid-activated char-coal, both supplemented with 10 g/L captan and 10 g/ L sucrose. Cuttings taken from C. atlantica and C. libani are difficult to root. Some rooting may occur on cuttings taken in late winter and treated with 8 g/L (8,000 ppm) IBA-talc (Dirr and Heuser 1987).

Cultivars of Cedrus species are more routinely propagated by grafting (Richards 1972; Lyon 1984; Blomme and Vanwezer 1986; Macdonald 1986; Dirr and Heuser 1987; Hartmann et al. 1990; Siniscalco 1995). Two reports have been published on the in vitro culture of C. deodara (Bhatnagar et al. 1983; Liu 1990). A method for in vitro propagation of C. libani through axillary bud production, a study of bud dormancy in vitro, and the detection of genetic variation of in vitro propagated clones have also been described (Piola and Rohr 1996; Piola et al. 1998, 1999).

ADAPTATION TO THE URBAN LANDSCAPE

The diversity of conifers available for planting in the urban landscape is quite large. There are many species and cultivars with numerous ornamental charac-teristics, such as dwarfed, weeping, and pyramidal, and there are a vast array of needle colors. The cedars are magnificent conifers that should not be overlooked for use in the landscape as a specimen plant or lawn tree, and in gardens or parks. Cedars are exquisite trees giving beauty to any area that can afford them a little space. The cedars offer resistance to climatic stress and good growth performance. They require abundant sunlight throughout their life and will thrive in almost any moist, deep, well-drained soil. They are unsuited for shallow, rocky soils where the plants either die as a result of drought when young or remain stunted if they survive. Cedars respond well to fertilizers, mulching, and supplemental watering during drought (Chaney 1993). All the cedars are difficult to transplant to some degree and are usually sold as container-grown plants.

Cedrus atlantica was introduced into cultivation around 1840 (Dirr 1990). A fast-growing species, the Atlas cedar prefers a sunny (or partial shade) location in a well-drained, moist, deep, loamy soil. It will tolerate sandy, clay soils if there is no standing moisture (Dirr 1990). Atlas cedar prefers an acid soil but withstands alkaline conditions (Dirr 1990). It grows well in the heat of the south and tolerates pollution and urban conditions better than C. libani or C. deodara. Cedrus atlantica should be planted in areas sheltered from strong, sweeping winds (Dirr 1990). It will suffer considerable needle burn and injury during cold winters when sited in windy locations.

Rot caused by Phellinus pini A. Ames can be a potential problem (Hansen and Lewis 1997), and the Deodar weevil occasionally damages the Atlas cedar, especially in the south. Cedrus deodara was introduced into cultivation in the United States in 1831 (Dirr 1990). The Deodar cedar prefers full sun, and a well-drained location is essential. The best growth is attained on deep, fertile soil, fairly porous, and in cooler locations (Tewari 1994). High atmospheric moisture is favorable (Vidakovic 1991). Cedrus deodara thrives on exposed sites, but young seedlings require side shade or light overhead cover to prevent desiccation (Tewari 1994). The Deodar cedar is affected by drought chiefly in the seedling stage (Tewari 1994). This species is generally wind firm and frost hardy (Tewari 1994), but young trees are prone to injury from frosts and cold wind. Peridermium witches'-broom caused by Peridermium cedri, and root and butt rots caused by Heterobasidion annosumBref. are particularly injurious to Deodar cedar (Hansen and Lewis 1997; Tewari 1994).

The Deodar weevil, Pissodes nemorensis Germar, can injure trees by feeding on young shoots of the crown, terminal killing, and branch-end flagging (USDAForest Service 1985). Cedrus libaniis a majestic tree introduced into cultivation during colonial times (Dirr 1990). Cedar of Lebanon prefers full sun and a deep, well-drained loamy soil. It is intolerant of shade and needs a pollution-free, dry atmosphere (Dirr 1990). Cedrus libani is quite winter-hardy, but grows rather slowly (Dirr 1990; Vidakovic 1991). Cedar of Lebanon is one of the most picturesque conifers in the landscape. Old specimens of Cedrus libaniwith their massive trunks, stiff habit, dark green foliage, and rigidly upright cones lend an aspect of dignity and beauty to parks and gardens (Dirr 1990; Maheshwari and Biswas 1970). There are no serious diseases or insect pests that affect Cedar of Lebanon. Cedrus brevifoliais a rare species seldom found in the urban landscape.

11

Medicinal and Economic Plants of Hilkot Watershed Area

INTRODUCTION

Hillkot watershed area is rich in a variety of medicinal and economic plants. Some of which are of pharmacopoeial importance while a large number are used in traditional system of medicine against several diseases by communities in primary health care system. Although this area possess good potential of medicinal and aromatic plants, still their richness is not known. This natural resource at present is not utilized according to resource availability, as a result a number of pharmacopoeial species became scare due to detrimental extraction by inhabitants. Besides, there are many other plants which are of substantial economic importance.

Pakistan Forest Institute, Peshawar is implementing a ICIMOD funded project entitled "People and Resource Dynamic in Mountain Watershed of Hindukush Himalayas in Pakistan." This project proposes an inter-disciplinary approach related to the management of natural resources and rehabilitation of degraded land thought community participation. Therefore, a preliminary rapid appraisal vegetation survey was carried out from 27 July to 3 August, 2000 to determine the floristic composition, present status and ethnobotanical uses of plants. The objectives of the study were:

- To find out floristic composition and present status of medicinal, economic plants and their market potential.

- Ethnobotanical study of plants used in folk medicines and other economic uses.
- To find out prospects of endemic medicinal and economic plant species that need conservation for sustainable utilization.
- To find out profitable drug species suitable for in-situ cultivation through community participation.

METHODOLOGY

- Keeping in mind the terms of reference (TORs), coniferous forests of the project area was visited to study floristic composition and medicinal plants.
- Meeting with the communities to get information about feasibility of growing fast selling drug species alongwith agricultural crops on community land to enhance their income.
- Various hamlets of the project area were visited to collect information with the help of informants about indigenous knowledge of local people about plants and their impact on the life of the people. In addition to short interviews with community members, discussion also took place with users and buyers of medicinal plants to know the existing marketing system and to suggest measure for improvement.

FLORISTIC COMPOSITION

The Hilkot watershed covers an area of about 20 km^2 with an latitudinal range from 1500 to 3200 m. Major land at foot-hill and moderate slope were brought under cultivation of cereal crops like maize and rice on terraces. The project area is thickly populated and characterized by grassland with rugged, steep to moderate slope. The southern aspects show the sign of degradation, while northern aspects are thickly vegetated. Ecologically the project area lies under moist temperate coniferous forests. The number of dominant species is small, in fact more or less pure or two species are admixtures. The conifer generally forms a fairly complete forest cover in the

Perai Reserved and in Guzara Forests. On moist depressions, deciduous broadleaved forest species of Populus ciliata (Poplar), P. deltoides (Poplar), Aesculus indica (Bankhor), Ulmus laevigata (Manu) are found alongwith under-growth shrubby vegetation. The natural conifer forests are met above 2000 m elevation at Perai and Guzara Forest from Saidan Gali to Tal Meteorological Station and Dabli (Naka-sher) to Hilkot.

The common trees are Pinus wallichiana (Kail), Cedrus deodora (Deodar), Picea smithiana and Abies pindrow. Among broadleaved trees; Quercus ilex Q. dilatata, Aesculus indica, Prunus, Ulmus, Corylus, Alnus, Morus, Litsaea and Euonymus were principal hardwood species and form local consociation with Indigofera, Lonicera, Rosa, Rubus, Desmodium, Viburnum constituting the under growth. Whereas on the farm lands Juglans, Ailanthus, Robinia, Diospyros, Zizyphus, Pyrus, Prunus were common, while along nullahs Alnus, Salix, Celtis, Ficus were conspicuous. Vegetation of Perai Reserved and Guzara Forests as well as on farm land area were analyzed and comprised of both natural and planted species given as under:

Trees

Abies pindrow (Fir), Aesculus indica (Bankhor, Ailanthus altissima (Ailanthus), Alnus nitida (Sharol), Broussonetia papyrifera (Jangli Tut), Cedrus deodara (Deodar), Celtis australis (Batkarar), Diospyros kaki (Wild amlok), Elaegnus hortensis (Singli), Ficus carica (Injeer), Ficus palmata (Ficus), Juglans regia (Akhrot), Melia azedarach (Bakain), Morus serrata (Tut), Pinus wallichiana (Kail), Platanus orientalis (Chinar), Populus deltoides (Poplar), Populus alba (Poplar), Populus ciliata (Poplar), Pyrus communis (Nashpati), Pyrus malus (Seb), Pyrus pashia (Batangi), Quercus incana (Rin), Rhus succedanea, Robinia pseudoacacia (Robinia), Salix tetrasperma (Bed), Taxus baccata (Barmi), Ulmus laevigata (Manu) and Zizyphus vulgaris (Unab).

Under-growth

Shrubs

Astragalus sp., Berberis lycium, Caragana ambigua, Daphne oleoides, Desmodium suvame, Elaegnus parviflora, Indigofera gerardiana, Lonicera sp., Onychium japonicum (fern), Plectranthus rugosus, Punica granatum, Sageretia filiformis, Spirea rosea, Sarcococa saligna, Viburnum nervosum and Zanthoxylum alatum.

Herbs

Acorus calamus, Angelica glauca, Arisaema tortousum, Artemisia sp., Bergenia ciliata, Cannabis sativa, Chenopodium album, Commelina sp., Corydalis stewartii, Epilobium roseum, Fragaria vesca, Geranium wallichianum, Heraceleum candicans, Hypericum dyeri, Launea nudicalus, Mentha longifolia, Nepeta catarica, Nepeta raphanorhiza, Nostrum officinale, Plantago major, Podophyllum hexandrum, Polygonum viviparum, Rumex hastatus, Scabiosa sp., Scutellaria chamaedrifolia, Silene vulgaris, Skimmia laureola, Solanum miniatuum, Valerina wallichii, Verbascum thapsus, Viola odorata and Urtica dioica.

Climbers

Hedrela helix, Jasminum grandiflorum, Rosa moschata and Rubus sp.

STATUS OF IMPORTANT ECONOMIC AND MEDICINAL PLANTS

Survey of important economic and medicinal plants conducted in different forest localities of Hilkot Watershed area indicated that a number of them having good demand can be collected in sizeable quantities, while others are threatened with extinction due to their heavy use by ever-increasing population pressure. Key plant species of economic and medicinal value given as under:

A. Plants growing wild can be collected for marketing

Acorus calamus (Bach), Arisaema tortosum (Sur-ganda), Cannabis sativa (Bhang), Chenopodium album (Bathu), Mentha longifolia (Jangli-podina), Nastrum officinale (Jangli-salad), Plantago major (Jangli Isabghol) and Verbascum thapsus (Gidar-tambaku).

B. Plants of commercial value can be cultivated on community land

Anethum graveolens (Ajmod), Asparagus adscendens (Musli-sufaid), Carum copticum (Ajwain), Foeniculum vulgare (Sonf), Hyoscyamus niger (Ajwain-khurasini), Linum usitatissimum (Alsi), Malva parviflora (Sonchal), Punica granatum (Annar), Saussurea costus (Kuth) and Vernonia anthelmintica (Kali zeri).

C. Threatened plants

Due to over-exploitation by locals resulted in the loss of plant population and need to be conserved. The threatened group of economic/medicinal plants categorized in three sub-groups.

i. *Endangered Species*: Distribution frequency of the following species is low in the visited localities and considered to be endangered with extinction. Angelica glauca (Chora), Aesculus indica (Bankhor), Celtis australis (Batkarar), Corydalis stewartii (Mamiri), Ficus palmata (Wild Injeer), Juglans regia (Akhrot) Paeonia emodi (Mamekh), Quercus sp. (Rien), Rheum emodi (Revand-chini) and Zizyphus vulgaris (Unab).

ii. *Rare Species*: These species have limited distribution in specific area. Their populations are decreasing in localized habitats and became endangered: Bergenia ciliata (Zakhm-e-hayat), Berberis lycium (Kashmal), Dactylorhiza hatagira (Salep-misri), Delphinium denudatum, Geranium wallichianum (Ratan-jot), Polygonum amplexicaule (Anj-bar), Skimmia laureola (Ner), Thymus serpyllum (Ban-ajwain), Valeriana jatamansi (Mushak-bala), Viola serpens (Banafsha) and Zanthoxylum armatum (Timar).

iii. *Extinct Species* : Elaeganus hortensis (Singli), Fraxinus excelsior (Sum), Fraxinus xanthoxyloides (Sum), Taxus baccata (Barmi) and Ulmus laevigata (Kain).

SILVICULTURAL CHARACTERISTIC OF KEY SPECIES

The silvicultural characteristics of key species have been elaborated in annex. Saplings of these key species may be raised either in project nursery at Hilkot/PFI or can be procured from Forest nurseries for planting in eroded area of watershed for conservation and rehabilitation.

MARKETING SYSTEM

Markets of Batal and Chatter-plain area are trade center of medicinal plants and black mushroom (Guchi). People collect a limited number and quantities of drug plants from all around the Hilkot valley during summer month. They at first dry it near home-yard and later sell them to local traders/shopkeepers. The seasonal traders purchase dried herbal drugs after grading from local shopkeepers, who then packed in gunny bags and forwarded to Mingora Market; the main trading centre in Swat where these commodities accumulate and ultimately find their way to down markets of the country for consumption and export. Different crude drug items sold by local dealers of Batal and Chatter-plain areas in the recent year.

ECONOMIC POTENTIAL FOR SUSTAINABLE HARVEST OF MEDICINAL/ECONOMIC PLANTS

Medicinal plant is a small component of Agriculture sector and contributes its share in economic development. The sustainable harvesting of plants having both medicinal and economic value has great potential. In fact, there is no local awareness about the proper collection of various species. Thus there is a need, therefore, to create awareness among local people of the importance of these plants and to provide them guidance and training in collection and processing to enhance their income.

ETHNOBOTANICAL OBSERVATIONS

A questionnaire was designed to document traditional knowledge of medicinal and other economic plants. Field studies were conducted in 3 hamlets and information obtained from Hakim and local people who had knowledge of therapeutic value of plants. These professionals belonged to different ethnic groups (Gujars, Swati and Syeds). Discussion of the properties of the plants were conducted with them with the help of informants. The number of peoples with whom the information were collected varied from 3 to 5 depending upon availability. Majority of peoples illiterate and about 15% of respondent were Primary, Matric and college education. More than 10 informants interviewed were active, cooperative and aged varied from 45 to 65 years.

The average number of house holds in the hamlet were 52 with a population of 570 heads, where people lived with their own way of life, believe and cultural heritage. The number of dependent per family heads were 10 but the average number of children were 6. Majority of the residents (60 to 90%) depend on farming.

Per capita land holding varied from 0.37 to 1.20 ha, which was not sufficient to fulfil the basic needs. They were doing other jobs in addition to farming. The annual income varied from 12,000 to 25,000 with an average of 18,500. Respondent benefitted from natural resources other than firewood like collection of medicinal plants, morel, grazing and grass cutting. They enhanced their income from the sale of non-timber forest products. Majority of the respondent reluctant to tell the exact quantity of medicinal plants and morel extracted in a season.

However, some of them (10%) told that 1 to 1.5 quintal (fresh wt.) of medicinal plants and morel were collected per family for sale and local uses in a season. The plant remedies usually employed in common ailments like cough, asthma, dyspepsia, skin diseases, typhoid, malaria, eye-trouble, body pain, cuts and wounds. Thus people saved handsome amount instead of purchasing costly allopathic medicines for the treatment of these diseases. The inhabitants only preferred to visit hospital/Doctors in serious cases. The information collected on traditional uses of medicinal and economic plants.

The use of plants is indicative of intimate dependence and relationship of the peoples of hilly areas with the vegetation in their vicinity. However, there is an ample traditional knowledge of medicinal plants which need to be fully documented for restoration of disappearing knowledge as cultural heritage can be used as a source of new medicine.

CONSERVATION AND CULTIVATION/PROPAGATION

Many medicinal plants like Paeonia emodi, Podophyllum hexandrum, Angelica glauca, Rheum emodi, Dactylorhiza hatagira, Geranium wallichianum, Polygonum amplexicaule, Skimmia laureola, Valeriana jatamansi, Viola serpens etc. and morels are collected every year by locals and no attempt is made for their conservation or to protect them against grazing pressure. If this practice continued, there is possibilities of complete exhaustion of a particular herb in near future. Therefore, it is necessary to establish demonstration plot in the farmer's field to promote cultivation of medicinal plants to uplift socio-economic condition of communities. Financial and technical assistance may be provided to the farmers of small holdings through PARDYP project for cultivation of fast selling drug species having constant demand in the market. These species will generate additional income to the communities.

CONCLUSION AND RECOMMENDATIONS

i. Preliminary rapid survey conducted in selected forest localities indicated that a number of valuable medicinal and economic plants became scarce due to ruthless extraction of their natural population and biotic interferences. It is imperative that a detailed survey of all commercially important medicinal and economic plants need to be carried out in Hilkot watershed area and suburbs in the next growth season to (i) assess their available potential in natural habitat, (ii) identify localities which abound in rich diversity of plants and (iii) preparation of a list of endangered plant species. This survey is essential to formulate a judicious plan of sustainable utilization, so that no species of medicinal or economic value of the area become endangered in future.

ii. The local populace needs guidance in identification, collection, conservation and cultivation of profitable drug plants with the help of Demonstration plot proposed to be established in the farmer's field. This training program can be provided by trained staff of PFI to create awareness and enhance professional capability of the community activists to maintain purity and quality of herbal drugs.

iii. Marketing survey of important herbal drug markets should be conducted to determine the supply and demand position and to identify the annual requirements of various plant-based manufacturing units. If this information is available, links with marketing channel between community organization and end-users will be established. Thus collectors and growers will be in a better position to get maximum benefit from the sale of herbal drugs.

iv. Nurseries/Drug farm of rare and profitable drug and economic plants should be developed by PARDY Project in collaboration with PFI for transfer of packages of cultivation practices along with seeds and planting stock to community for pilot scale cultivation. It should also prove beneficial to the entire community if female members are encouraged to become involved in the cultivation of medicinal plants as herbal gardening. Thus in-situ cultivation will boost up the mass production of fast selling drug species and their sustenance in natural habitats.

v. Phyto-chemical analysis should be carried out with the help of HEJ Research Institute of Chemistry, Karachi to find out active principles present in medicinal plants which have a potential for export or as concentrates. In case of positive result, market should be identified with the help of exporters. Any financial benefits which accrue from such activities would substantially benefit local communities.

vi. Project nursery established at Hilkot is in sufficient to meets the local requirement of planting stock. It should be extended and key species of conifers like Abies pindrow, Cedrus deodara, Picea smithiana Pinus wallichiana and Taxus baccata may be raised alongwith broadleaved species viz: Aesculus indica, Ailanthus altissima, Amorpha fruiticosa, Alnus nitida, Celtis australis, Elaeganus hortensis, Ficus palmata, Ficus carica, Fraxinus excelsior, F. xanthoxyloides, Gleditschia tricanthos, Juglans regia, Morus serrata, Platanus orientalis, Populus ciliata, Populus deltoides, Quercus dilatata, Q.ilex, Robinia pseudoacacia, Sapindus mukorosii, Ulmus laevigata and Zizyphus vulgaris. These species may be utilized for rehabilitation of eroded land and conservation of plants to restore species diversity.

vi. Planting of Eucalyptus and Chir seedlings in moist temperate forests give negative impression to the visitors as these species are not suited to the climatic conditions.

SILVICULTURAL CHARACTERISTIC OF KEY SPECIES

Aesculus indica (Bankhor)

It is a native species and found in Swat, Hazara, Murree hills and Azad Kashmir. It grows on deep, rich, moist soils, but prefers moist shady sites. It requires moist temperate climate with a temperature range of -20 to 35^0 C and at an elevation range of 1200 to 3300 m. It is an important component of the coniferous forest. It can be propagated either by seed or vegetative means. Fresh seed give better germination as it has low viability.

Celtis australis (Batkarar)

It is a moderately fast growing, medium to large sized deciduous tree, flowers between February and May and fruit matures in April to September. It grows on a variety of soils in a sub-humid, cool to semi-arid, warm sub-tropical winter/ monsoon climate, usually at elevations of 400 to 2000 m with a temperature -10 to 53^0 C. Its fodder is highly palatable and is used for forage in the sub-mountainous regions. It is reproduced from cuttings, coppice, and seed. Seeds have high viability.

Elaeganus hortensis (Singli)

It is a medium fast growing shrub or small tree and grows well within a temperature range of -10 to 35^0 C in moist temperate climate. Fruit edible and matures from July to August. It can be reproduced from seed.

Ficus palmata (Jangli injeer)

It is fast growing tree and grows well within a temperature range of -10 to 350 C. It flowers in April and May. Seeds are very small. Fruit matures from October to November. It can be propagated both from seed and by vegetative means. It will also coppice.

Fraxinus excelsior (Ash, sum)

It is a medium fast growing, deciduous tree, found from Swat to Azad Kashmir, Hazara and Balochistan. It flowers between April and May and seed matures between May and October. It prefers a humid cool temperature, sub-tropical winter/ monsoon with a temperature range of -2 to 35^0 C at elevations between 1200 and 2700 m. It can be propagated from seed and by vegetative means and coppices easily.

Fraxinus xanthoxyloides (Ash, Sum)

It is a shrub or small deciduous tree found in Gilgit Agency, Chitral, Dir, Swat, Hazara, Kurram, and Balochistan. The flower blooms between March and April and fruit is a winged seed, that mature between May and August. It prefers an arid to semi-arid, cool to cold temperature, Mediterranean climate with a temperature range of -20 to 35^0 C at elevations between 1000 and 2500 m. It coppices easily and reproduced both from seed and by vegetative means.

Gleditschia triacanthos (Dozakh, Honey Locust)

The tree is native to the U.S.A. but has been successfully planted in Africa, Australia, New Zealand, South America, Pakistan and other countries of the world. It is a deciduous tree with thorny branches.

It flowers between May and June, The fruit matures between September and August. It prefers an arid to semi-arid, cool to cold temperature, Mediterranean climate with a temperature range of -2 to 35^0 C at elevations between 1000 and 2500 m. It is frost and drought resistant. It can be produced both from seed and by vegetative means. Under cold storage seed will remain viable for 2 years. Hot water or acid treatment is need to overcome seed coat dormancy.

Juglans regia (Akhrot, Walnut)

It is a deciduous, medium fast growing tree found in Northern Areas, Dir, Swat, Hazara, Murree and Azad Kashmir. It flowers between February -April while fruit matures between June to October. It prefers a humid, cool temperate climate, usually from plains to 2300 m with a temperature range of -10 to 35^0 C. It coppices at an early age. This is a very valuable tree of coniferous forest but planting program may be needed to preserve it and insure its position as a member of the coniferous forests. Because of its fruit high valuable wood and bark, it could be a part of farm forestry. It is generally produced from seed but can also be propagated by vegetative means.

Morus serrata (Tut, Mulberry)

It is a large, fast growing deciduous tree. It requires a precipitation zone of 750 to 1250 mm/yr or more, and prefers a semi-arid, cool to cold subalpine temperate, sub-tropical winter/ monsoon, climate with a temperature range of -10 to 40^0 Cat elevations up to 3300 m. It coppice easily, can be grown in mixed stands. It can be propagated both from seed and by vegetative means.

Populus ciliata (Palach, Himalayan poplar)

It is dioecious fast growing tree grows in Azad Kashmir, Northern Areas, Gilgit, Murree Hills, Hazara, Swat, Dir, Chitral and Tirah. It grows best on deep moist soils, but can grown on a variety of sites including rocky exposed land slides areas. It prefers a humid, semi-arid cool, cold temperate, climate with a temperature range of -20 to 35^0 C. It is a component of the coniferous forest and can be easily regenerated by cuttings. It has the potential to control erosion on stream bank, steep hillsides and should be planted in problematic watersheds.

Populus deltoides (Sufeda, poplar, Northern Cottonwood)

It is a large, deciduous fast growing tree native to North America. In Pakistan it has been successfully planted in the plains as well as in the hills. It prefers a semi-humid to semi-cool, sub-tropical temperate climate in a temperature range of -20 to 35^0 C. It does coppice and propagated by vegetative means. It is suitable species for erosion control and reforestation in watershed area.

Quercus dilatata (Barungi)

It is a large evergreen tree found in the Himalayas mountains specifically in Dir, Chitral, Swat, Hazara, Tirah, Kurram Agency, Murree Hills, and Azad Kashmir. It requires a humiu to sub-humid, cool, temperate climate with a temperature range of -20 to 35^0 c at elevations between 1600 and 2900 m. It will coppice readily. It can be propagated both from seed and by vegetative means. The fruit is a nut and has a very low viability. This tree is an important component of the coniferous forest. In order to maintain diversity attempts must be made to ensure its regeneration.

Salix babilonica (Bed-e-majnu, Weeping Willow)

A large to medium sized, relatively fast growing, deciduous tree found in the Himalayas valleys along water courses specific to Azad Kashmir, Kaghan, and Swat. It prefers a sub-humid, semi-arid, sub-tropical monsoon climate with a temperature range of -2 to 40^0 C. It does coppice. It can be produced from seed and by vegetative means. However most tree result from root suckers or cuttings. It is a good tree to plant on moist denuded hills and gullies areas to check erosion.

Sapium sebiferum (Charbi and Chinese Tallow tree)

It is moderately fast growing, medium sized deciduous tree grow naturally along stream beds in Abbottabad and suburbs area. It prefers a sub-humid to semi-arid, cool-warm subtropical monsoon climate with a temperature range of -10 to 40^0 C up to an elevation of 2000 m. It is propagated by vegetative means and suitable for planting in watershed areas.

Taxux baccata (Barmi)

It an important companion of Kail and Fir conifers species in moist temperate forests. It is an evergreen, slow growing tree which prefers to grow under partial shade. It can be propagated through seed and vegetative means.

Ulmus laevigata (Manu)

It is a moderately fast growing, large sized, deciduous tree found in Swat, Hazara, Murree, Azad Kashmir and Balochistan. It grows on deep, well drained soils and prefers moist ravines. It prefers a sub-humid, cool to semi-arid, warm sub-tropical temperate climate, usually at elevations of 1000 to 3000 m with a temperature range of -10 to 35^0 C. The fodder is highly palatable and is used for forage in the mountainous regions. It has potential for soil erosion control. It can be propagated from seed and seeds are small and lose viability shortly after ripening.

Zizyphus vulgaris (Unab)

It is moderately fast growing, deciduous small tree with straight bole found in Swat, Hazara, Murree, Azad Kashmir and Balochistan. It flowers between May to June and fruit matures between July to August. It grows on deep, well drained soils and prefers moist ravines. It prefers a sub-humid, cool to semi-arid, warm sub-tropical temperate climate, usually at elevations of 1000 to 3000 m with a temperature range of -10 to 35^0 C. The fodder is highly palatable and is used for forage in the mountainous regions. It has potential for soil erosion control. It is reproduced from seed and cuttings.

RAISING TREES AND SHRUBS FROM SEED

Growing a few trees from seed is a relatively easy and extremely rewarding hobby. But without previous plant production experience, it is unwise to expect to grow even a few hundred plants, for example, for a recently acquired piece of land. And considerable caution and detailed planning are needed before setting up a commercial tree nursery.

This Practice Guide is aimed principally at people interested in raising small numbers of trees-to plant in the garden, raise for bonsai, or perhaps just for the fun of finding out what it's like to grow a few trees from seed. Nevertheless, in describing some of the basic properties of tree seeds, principles of seed handling, plus explaining how to avoid or overcome common problems, this Guide will be a useful reference for all. It covers approximately 120 woody species which are either native to, or commonly grown in, the British Isles.

Understanding Tree Seeds

Most seeds appear drab, dried, apparently lifeless objects. They are consequently often treated as though they are inert matter, but they are living organisms. Handled with care they will fulfil their potential to germinate, grow and develop into healthy plants, whereas mishandling will damage or kill them.

Tree seeds (in comparison to flower, vegetable and agricultural seeds) are some of the most fragile but also interesting of all seeds. Some of their characteristics are unique and they also have several quirks and idiosyncrasies that make them the eccentrics of the seed world, as described in this section.

Erratic Seeding

One oddity is the erratic and unpredictable fruiting of trees. In some years, trees produce very large quantities of cones, fruits or seeds and in other years almost none. There are some people who think that the intermittent production of large seed crops (so-called 'masting') is a distinct ecological advantage. They propose that 'barren' years cause predator populations to fall, and in mast years there are more seeds than all the surviving predators can possibly eat, so a larger proportion of seeds will escape predation and sprout the next spring. However, the less romantic explanation is that inconsistent seed production merely reflects annual variations in weather. And presumably, in the context of such a long-lived entity as a tree, erratic seed production is of little evolutionary consequence. But, whatever the explanation for masting, it is important to appreciate the phenomenon. Do not expect to be able to collect a guaranteed crop of tree seeds every year. Small crops are the norm, large crops the exception. The general rule is, when it looks like it is going to be a good seed year-plan to collect.

Unproductive Seeds

A second peculiarity of nearly all trees is that, in addition to live seed, they habitually produce a large proportion of empty and dead seeds. From the human perspective, this also seems very odd that an organism invests time, effort and resources into making something unproductive. As with masting, it has been suggested that this is a special survival strategy. For example, it is believed that, firstly, the tree benefits in that it produces empty, decoy seeds with minimal energy expenditure; and secondly, that potential predators will learn that they waste too much time and effort in sorting through empty and dead seeds which will act as a deterrent and encourage them to seek food elsewhere.

Again however, it is more likely that in the context of a large and long-lived tree, a high proportion of empty and dead seed is simply not enough of an evolutionary disadvantage to have become selected out. Nevertheless, it is particularly important to bear this characteristic in mind with the previous point on erratic seeding, because there is often a much higher proportion of dead and empty seeds in a poor crop year than a good crop year-so it is doubly advantageous to collect tree seeds in mast years.

'Suicidal' Seeds

Another curiosity is that some very small seeds, such as willow and poplar, and some very large fruits, such as oak, sycamore, sweet chestnut and horse chestnut, die quite soon after being shed from the tree-one of the last properties you would normally associate with seeds. The fruits are killed if they dry out and at present there is no known method of doing anything more than slowing down their rate of deterioration. It is therefore only worth collecting seeds of these species if you can sow them fairly quickly, or are prepared to suffer significant losses over, for example, one winter's storage. Seed longevity and associated storage characteristics are considered more fully under the headings 'orthodox', 'recalcitrant' and 'intermediate'.

'Malingering' Seeds

Finally, there is a phenomenon known as dormancy. There are many definitions of dormancy, but for anyone interested in raising trees from seed all you really need to know is that the seeds of some species have an absolute requirement to be 'pretreated' before they will germinate at all; and the seeds of other species are merely reluctant to germinate until they have been pretreated. Further information on 'dormancy', 'hard-seededness' and 'pretreatment'. As these examples of a few unusual tree seed characteristics show, never expect every tree seed sown to produce a seedling. A three-hundred-year-old English adage sagely predicts 'One to sow, one to grow, one for the mouse and one for the crow', meaning that you are only likely to get one tree seedling for every four seeds sown. It is therefore a sobering thought that all the trees of the future are in just a few seeds of today. The rest of this Practice Guide will consider the process of raising trees and shrubs from seed.

Flowering and Fruit Development

Every potential plant propagator should know something about the reproduction of trees. In common with most higher organisms, it is generally believed that it is an advantage for trees to promote gene mixing through the transfer of pollen between different plants (so-called 'out-breeding'). A corollary of this is that there are mechanisms to prevent or at least reduce the chances of 'in-breeding'. A few trees achieve this by having separate sexes-they exist as male and female individuals.

This makes self-pollination ('selfing') impossible. Other species produce male and female flowers on the same tree but on different branches, or at different times, or both. This again provides a physical and/or temporal obstacle to selfing. However, the flowers of the majority of trees are hermaphrodite-that is they contain both male and female structures. These flowers use several anatomical, temporal and biochemical mechanisms to reduce the chances of selfing. However, this means that some thought should be given to the different self-incompatibility systems of different species and the effect that collecting in different situations and for different purposes might have on the likely gene-pool of the seeds.

Holly is often dioecious-a) shows swollen, central carpel and four, reduced anthers of the female flower; b) shows reduced, central carpel and four more prominent anthers of the male flower. As the most obvious example, do not expect to get fruits from the male individuals of species that are usually dioecious (e.g. ash, holly). It is also relevant to consider pollen source. For example, whether the pollen is mainly local or whether it has been blown in from a distance. This may be especially important if your aim is to collect seeds of native trees and shrubs, from a local source, undiluted by genes from afar.

Another consideration is that self-incompatibility systems are not fool-proof, and their failure is frequently the explanation for why trees produce such large percentages of empty and dead seeds. Pollination, fertilisation and early fruit development occur, and it is only later that embryo abortion and possibly tissue resorption takes place. As a general rule, an identifiable group or defined population of at least 20-30 cross-pollinating individual trees will help to ensure that a reasonable amount of inter-breeding has taken place and that the gene pool of the seeds is not too narrow.

There is one other genetic consideration regarding the likelihood of climate change. A number of groups currently encourage the collection and propagation of native trees from local seed sources-their aim being to preserve locally adapted genotypes. However, if the UK climate is likely to significantly change in the next 50-100 years then perhaps seed sources adapted to the local environment may not be so appropriate. Finally, to emphasise two points made in the previous section. Firstly, most trees do not produce a good fruit/seed/cone crop every year, and secondly, in a poor crop year, seed quality tends to be lower than in a good crop year. It is therefore wise to monitor flowering and the development of fruit quality and quantity and only collect in years when there appears to be a worthwhile crop.

Collecting, Handling and Processing

One major practical difficulty about collecting tree seeds is how to collect from such tall plants. In the case of large fruits and seeds which fall directly to the ground (e.g. acorns, sweet chestnut and horse chestnut) the solution can be to gather the seeds from beneath the tree, or for larger collections, possibly clear any ground cover and spread out nets or tarpaulins. This way you can also be fairly sure that, since natural dispersal has taken place, the seeds are likely to be fully mature.

But, if animals or birds are likely to get to the fruits before you (e.g. cherries, rowan and hazel) or if seeds are wind-dispersed and either tiny (e.g. alders, birches, poplars, willows and most conifers) or medium sized (e.g. ash and sycamore) then the only safe option is to identify trees with accessible branches and pick the fruits just before maturity. Two other things should be considered.

If the trees are on private land it is essential to obtain permission from the owner(s) before collecting any seed. Secondly, always remember to leave some seeds behind-they are a vital food source for wildlife. For collection, handling and processing purposes, tree fruits/seeds and cones can be split into the following groups with similar properties: legume pods, large 'dry' single fruits, medium sized 'dry' fruit clusters, medium size 'dry' fruits containing many seeds, fleshy fruits containing one or more seeds.

Legume Pods

Examples include acacia, broom, false acacia, gorse, laburnum (pictured) and tree lupin Legume pods should be picked from the tree/shrub when they turn from green to brownish-black, and before they have fully opened and shed their seeds. The pods should be spread out to air dry and then placed in a bag and squeezed, beaten or threshed to release the seeds. The crushed, dried pods can then be separated from the seeds by sieving, fanning, or a combination of techniques. Note that the pods and seeds of laburnum are considered poisonous if eaten.

Large 'Dry' Single Fruits

Examples include beech (pictured), chestnuts, hazel, oaks and walnuts Acorns and chestnuts are easiest to collect from the ground. Small quantities can be gathered by hand; larger quantities can be swept or raked up. Beech nuts are a little bit smaller and can either be collected from accessible branches when they have turned brown, or from the ground. However, with beech it is advisable to avoid the first nuts to fall, as these are prone to have a particularly high proportion that are empty. Hazel nuts and walnuts must be collected from the tree and the development and maturation of the crop will have to be monitored very closely or squirrels may get there first. The most processing that these large, dry fruits ever need is perhaps some cleaning to remove larger debris such as twigs, leaves and the 'pixie pipes' of acorns (the cupules plus fruit stalk). It is also important to take great care not to expose acorns and chestnuts to drying

Medium Sized 'Dry' Fruit Clusters

Examples include ash (pictured), maples, limes and hornbeam Collect clusters of ash fruits (commonly called 'keys'), maple fruits (commonly called 'helicopters'), limes and hornbeams from the tree as they turn from green to brown. Strip individual fruits from bunches and remove the stalks to reduce bulk and facilitate subsequent handling.

Medium Sized 'Dry' Fruits Containing many Tiny Seeds

Examples include alder and birch strobiles and conifer cones (pictured) The botanical term for the structure containing the fruits of alders and birches is 'strobile', but in common with the seed-containing structures of many conifers they are often loosely referred to as 'cones'. The cones of birches and most cedars and firs are prone to disintegrate when fully mature so they must be collected from the tree before this happens or all the seeds will be lost. In comparison, the ripe 'cones' of alders, Lawson's cypress and most pines and spruces open and close in response to dry and wet weather as they are 'programmed' to shed their seeds over several days, weeks or even months.

It is therefore possible to miss the first seed shed of these species, yet still collect seed. Place the cones of either 'disintegrating' or 'opening and closing' cones in a paper bag at room temperature and allow them to dry. The former group will disintegrate to leave a mixture of winged 'seeds' and 'bracts'. However, there is often so little difference between the sizes, shapes or weights of the different components that the mixture is usually too difficult to separate. The latter however, will open, release their seeds and the intact cones are either easily removed by hand, or the seeds can be separated by blowing.

Fleshy Fruits Containing one or more Seeds

Examples include apple, cherry, hawthorn (pictured), holly, juniper, mulberry, privet, rose, yew Fleshy fruits are those that are soft, squidgy and frequently sticky. They are also often the most difficult to judge when to collect. They will either need to be carefully hand-picked, avoiding thorns where necessary, or possibly shaken from the branches. Apple seeds are an example of one of the difficulties of using the external appearance of a fleshy fruit to judge the internal development of the seeds. It is usually necessary to pick a few 'pomes' when they first turn from green to yellow/orange and cut open the centre. If the seeds are still white, or only partially brown then wait and re-sample weekly until the surface of the seeds is uniformly brown before progressing to the final collection. Cherries exhibit several difficulties. Firstly, you will generally need to collect them before they are fully ripe-to beat the birds! Secondly, it is important to remember that different cherry species ripen at different times of the year, and also that they ripen to different colours. For example, wild cherry (Prunus avium) ripens from green/yellow to red in June/July; whereas sloes (Prunus spinosa) turn blue/black in September/October.

Collecting yew berries has its own set of problems. It is vital to remember that, although the red flesh of the 'berry' surrounding the seed is regarded as safe to handle, all other parts of the tree, including the seeds, are toxic and may cause skin irritation. It is therefore necessary to exercise reasonable caution when collecting and handling these fruits, for example by wearing rubber gloves and washing hands thoroughly after processing. Fleshy fruits are also some of the most awkward and certainly the messiest to process.

Fruits such as apple, cherry, hawthorn, holly, juniper, mulberry, privet, rose and yew, usually need highly individual combinations of maceration (or controllled mashing), extraction, cleaning, soaking and washing procedures to separate and clean the seeds. The first step is generally very gentle maceration or even squeezing of the sticky fruits by hand to release the harder coated fruits or seeds within. Sometimes this may be assisted with a water soak and very occasionally a little fermentation can help.

However, for seeds such as hawthorn, holly and rowan, fermentation can be significantly harmful or even fatal and is therefore to be avoided. Subsequently, most seeds will need repeated washing not only to remove the clinging remnants of sticky flesh, but also as a means of removing chemicals that have the potential to inhibit germination. A large sink plus draining board, rubber gloves, several buckets, bowls and assorted sieves are required when processing fleshy fruits. Throughout seed collection and processing it is essential to keep a record of the identity of each seed lot. Minimum requirements to note are species, place and date of collection. Details of how and when the fruits/seeds were temporarily stored, transported and subjected to different processing and cleaning procedures will help improve next year's techniques.

Following points contains a checklist of things to consider before starting seed collection.

Questions to consider before starting seed collection:

o Is permission to collect necessary?

o Are you sure of tree identification?

o Does this species fruit regularly or intermittently?

- o What is approximate date of fruit ripening?
- o Are fruits prone to sudden dispersal or early predation?
- o Is collection of fallen fruits from the ground possible?
- o Is hand picking from branches possible?
- o Is an extendable pole-pruner or pole-saw an option?
- o What is best container for collection and possibly temporary storage, e.g. bucket, polythene bag, paper bag, sack, open tray, lunch box?
- o What are best labelling materials, e.g. pencil and paper, plastic plant labels and indelible pen?
- o What is minimum information to record, e.g. species, place and date of collection?
- o What conditions are best for temporary storage (before cleaning or processing), e.g. airtight, well ventilated, cool, refrigerated?

At the end of fruit/seed processing a decision will have to be made on storage. This will depend partly on whether the seeds can be stored or not, partly on whether storage is necessary, and partly on how long they will need to be stored.

The next section provides some guidance on the storage characteristics and likely longevity of the seeds of different species. (The widespread usage of the term 'dry', to describe fruits that are not 'fleshy' 'squidgy' or 'sticky', is rather unfortunate in relation to seed storage characteristics: for example, acorns have been described as large, 'dry' fruits, yet it will be seen in the next section that excess drying with kill them.)

Seed Storage

The seeds of most plants are easy to store (so-called 'orthodox'). The seeds of a small minority of plants are currently impossible to store (termed 'recalcitrant'). And in recent years it has become increasingly widely acknowledged that there are seeds between these two extremes (termed 'intermediate'). In other words, there is a range of seed storage characteristics.

The ease or otherwise of seed storage generally depends on whether the seeds of particular species can be dried or not. Seeds of recalcitrant species at a low moisture content are likely to be dead. Orthodox seeds at a high moisture content will need drying before storage. It is therefore very useful to be able to measure the percentage moisture content of seeds.

Most tree seeds can be stored for longer at lower moisture contents (6-8% moisture content for most pines and spruces, 10-15% moisture content for most firs, cedars, cypresses and broadleaves). Recalcitrant seeds, for example, oak, sweet chestnut, horse chestnut and sycamore, begin to die below 40% moisture content. Seed moisture content is generally reported as the weight of water, expressed as a percentage of the fresh weight of the seeds. This is called the moisture content on a fresh weight basis.

Note: moisture content can also be calculated on a dry weight basis, but this method is less common and is not used here.

Moisture content measurement is generally carried out on a minimum sample weight of 5 g. However, it has also been shown that the minimum number of seeds in a moisture content sample should never be below 30 seeds. Method A describes how to measure seed moisture content for small-seeded species-those containing more than 30 seeds per 5 g, and Method B for large-seeded species-those which would contain fewer than 30 seeds per 5 g.

Method A-Smaller seeded species, where a 5 g sample exceeds 30 seeds

1. Accurately weigh a 5 g sample of seeds (this is the 'fresh weight').
2. Dry the sample (preferably at approximately +105°C for between 18-24 hours or until two consecutive weighings give a constant weight).
3. Take a final 'dry weight'.
4. See below for calculation.

Method B-Larger seeded species e.g. acorns, chestnuts and sycamore:

1. Take 30 clean, firm, healthy and apparently live fruits.
2. Cut the fruits into approximately 3 x 3 mm cubes and mix the sample as quickly as possible to minimise drying.
3. Accurately weigh a 10 g sub-sample of the cut fruits (this is the 'fresh weight').
4. Dry the sub-sample (preferably at approximately +105°C for between 18-24 hours or until two consecutive weighings give a constant weight).
5. Take a final 'dry weight'.
6. See below for calculation.

Calculation for Method A and Method B

Assuming that any reduction in weight reflects moisture loss:

Moisture content (as % fresh weight) = (fresh weight [g]-dry weight [g]) x 100 fresh weight [g]

Note: for fruits containing significant amounts of volatile substances such as resins, fat, and oils, this is slightly erroneous because these substances are also driven off by heating. However, even for these species the weight of volatile substances is usually relatively small, and does not significantly affect the moisture content calculated.

Orthodox Seeds

At one end of the range of storage conditions are the seeds which can be dried without harm. The cells of these seeds can not only be dried, but (once dry) they can then be frozen, stored and relatively easily revived.

Fortunately, these seeds are in the majority and that is why they are called 'orthodox'. In general, most pine and spruce seeds are orthodox and can be safely dried to 6-8% moisture content (fresh weight basis). Whereas the seeds of most orthodox broadleaved species are best dried to only 10-15% moisture content.

In Britain, when the freshly collected seeds of these two groups of species are left in a shaded, cool, well-ventilated room, they ultimately tend to equilibrate at approximately these moisture contents. Nevertheless, once dried it is preferable to avoid moisture content fluctuations by storing orthodox seeds in airtight, waterproof, sealed containers-a tightly tied, thick polythene freezer bag is ideal. At these moisture contents, and a storage temperature of +3°C to +5°C (the main compartment of a refrigerator) most 'orthodox' tree seeds will exhibit little deterioration over 5-15 years. In addition, because the cells of these dried seeds do not contain enough water to freeze solid, longevity is further increased by reducing the storage temperature to-10°C to -20°C (the icebox of a refrigerator or a deep freezer).

There are two important guidelines about the effects of storage conditions on the longevity of 'orthodox' seeds. Firstly, a 1% increase in moisture content will halve the storage life (at any given temperature). Secondly, a 5°C rise in storage temperature will halve the storage life (assuming no changes to moisture content). These dramatic effects emphasise why it is so important to avoid moisture content and temperature fluctuations during seed storage.

Recalcitrant Seeds

At the other end of the range of storage conditions, the seeds of a small minority of species cannot be dried, cannot be frozen and are therefore highly perishable. They have been named 'recalcitrant' seeds. The cells that make up the seeds of these species are like the cells of most other animal and vegetable tissues-if they are frozen or dried, they die. The species in this group are given in Table 1. Optimum storage conditions for freshly collected, temperate, recalcitrant seeds combine low (but not freezing) temperatures (+3°C to +5°C) to slow seed deterioration and minimise fungal growth, with high humidity to retard drying. But even under these conditions, the seeds of some poplar and willow species are extremely recalcitrant and all die within days of dispersal or collection. Acorns and chestnuts kept under similar conditions usually suffer 60-70% losses over a couple of years. Practical guidance on how best to handle recalcitrant seeds is given on pages 16-18, and in Forestry Commission Practice Note Handling and storing acorns & chestnuts and sycamore fruits.

Intermediate Seeds

In between the two extremes of seeds with either 'recalcitrant' or 'orthodox' storage characteristics are seeds with so-called 'intermediate' properties (see Table 1). These seeds can usually be dried to 10-15% moisture content without harm, but if further drying is possible, it must be done much more slowly and carefully. At the storage temperatures most commonly used for orthodox seeds (i.e. +3°C to +5°C) they also tend to deteriorate more quickly. Hence, for practical purposes all seeds with intermediate storage characteristics should be dried to 10-15% moisture content and then stored at -10°C to -20°C. Having used Table 1 to identify whether seeds of a particular species can be stored it must then be decided if it is necessary to store them or not. This will not only depend upon when it is intended to sow the seeds, but also the seeds dormancy characteristics and whether there is enough time available to dry and store them before a dormancy breakage 'pretreatment' should begin.

HARD-SEEDEDNESS, SEED DORMANCY AND PRETREATMENT

In many naturally reproducing plant populations a delay between seed shed and subsequent germination can be a significant ecological advantage. Postponing emergence clearly allows extra time for dispersal (whether by wind, water, birds or animals) to potentially greater distances. In any region with seasonal variations, it has the additional advantage that the germination of at least some seeds will coincide with the most favourable season for seedling growth and establishment.

Natural selection has therefore favoured many species with seeds which do not germinate immediately after dispersal-even when the environmental conditions are ideal (e.g. ample water, a suitable temperature for growth and good aeration). When seeds are inactive because they are 'immature' or possess a layer of tissues that prevent water uptake or gaseous exchange, they are said to be 'quiescent'. But when seeds can be shown to be fully hydrated and alive yet still remain outwardly inactive-they are said to be 'dormant'. It is likely that passage through a bird gut removes germination inhibitors from the fruit skin and pulp, however, intact seeds still retain their innate 'deep-dormancy'.

A dormant seed is one that is known to be alive, but remains outwardly inactive and does not germinate when we expect it to. The seeds of some species remain dormant under all conditions, and the seeds of others are only dormant under some conditions. There are a number of natural processes, seasonal stimuli or combinations of the two which overcome dormancy and stimulate germination. And there are a number of artificial substitutes which humans can use to mimic such 'dormancy breakage'. These are called pre-(sowing) treatments or pretreatments for short. Nearly all tree seeds exhibit dormancy of one sort or another and therefore fast, uniform and complete germination is only possible after live, dormant seeds have been exposed to a natural or artificial dormancy breakage pretreatment.

Hard-seededness

Temperate legume tree seeds (e.g. broom, false acacia, gorse, laburnum, tree lupin) possess a seed coat which prevents or at least significantly retards the germination of otherwise live seeds. This is usually caused by a layer (or layers) of tissue that acts as an impermeable barrier to water uptake, respiratory gas exchange or both. Hence the embryo tissues are either unable to hydrate or respire or both. There is no fundamental, physiological block to embryo growth, merely a physical barrier which excludes one or more of the essentials for growth from the tissues. This condition is known as 'hard-seededness' and because it is an in-built physical barrier to germination, some people do not acknowledge that it is a form of dormancy-but the result is the same: poor or non-existent germination, until the seed coat is 'pretreated'.

Seed Dormancy

In contrast to hard-seeded species (which do not germinate because the outer layers prevent water uptake or respiration) the seed coats of most tree species are permeable to water and respiratory gases. As a consequence, in favourable conditions for growth, the embryo tissues absorb water, metabolise and respire. Yet even when environmental conditions are ideal (e.g. ample water, a suitable temperature for growth and good aeration) most freshly collected (and even dry-stored) seeds still remain outwardly inactive and exhibit no signs of germination. Seeds with these characteristics exhibit true, physiological 'dormancy'. The seeds of some species remain outwardly inactive under all conditions and are said to be 'deeply dormant'. The seeds of other species only remain inactive under some conditions and are therefore described as 'shallowly dormant'. All dormant seeds require some form of pretreatment before they acquire their full ability to germinate.

Pretreatment

In nature, the delayed germination of many 'hard-seed coated' species is overcome in several ways. For example, animal teeth can fracture the coat; bird or animal intestines can mechanically abrade, chemically digest and microbially breakdown the seed coat; and fire, repeated freezing and thawing, and wetting/drying cycles; and fungal and microbial breakdown in the soil can all 'scarify' a hard seed coat. Each of these processes will allow subsequent hydration of the embryo and thus stimulate germination, providing it does not harm the embryo within. In comparison, physiologically dormant seeds tend to respond to other natural environmental stimuli. For example, they can perceive:

- o transitions between one or more cold winters, followed by warmer springs;
- o cycles of burial followed by re-exposure to light;
- o sequences, or combinations of these changes.

These environmental changes act as natural dormancy breakage mechanisms and they can be reproduced in artificial dormancy breakage pretreatments. For example, various physical, chemical, biochemical and even biological methods can be used to pretreat hard-seeded species. Also deliberate exposure of seeds to seasonal temperature changes or thermostatically controlled equipment can be used to artificially pretreat deeply-and shallowly-dormant species.

INTERACTION BETWEEN STORAGE PROPERTIES AND DORMANCY CHARACTERISTICS

The previous two sections of this Guide show that storage and dormancy characteristics are the two properties most likely to determine tree seed handling and propagation techniques. For example, in (approximately) ascending order of overall difficulty of germination:

1. Leguminous trees and shrubs like laburnum, broom and gorse are very easy to dry, handle and store. The only special consideration is that the seed coat must be 'scarified' (i.e. abraded, punctured or softened) to permit germination.
2. Alders, birches and most conifers are also easy to dry, handle and store, but their shallow dormancy characteristics can make them a little more challenging to germinate, especially to achieve the maximum germination possible.
3. Some firs, cedars, cypresses and a few other conifers are slightly more difficult again to germinate. They can be dried and stored but often die within weeks at room temperature and ideally require sub-zero storage temperatures. They also exhibit shallow dormancy and there is some evidence that some seed lots can suffer damage during water uptake.
4. Poplars, willows, oaks, sweet chestnut and horse chestnut are the least dormant and easiest to germinate of all tree species, but they are all very awkward to handle, and simply impossible to store because they cannot tolerate drying and are therefore highly perishable.

Orthodox

Some poplars and willows will deteriorate within hours and die over a few days. As a general rule, 10-50% of oak and chestnut populations will deteriorate between autumn collection and spring sowing.

5. Most broadleaved trees (plus junipers and yews) are relatively easy to dry, handle and store, but their deep dormancy makes them significantly more difficult to germinate because pretreatments are not reliable and pretreatment times are so lengthy.
6. A few broadleaves (e.g. beech, laurel and Norway maple) combine the intermediate storage characteristics with deep dormancy. This means that they require careful drying, relatively precise storage conditions and also very lengthy pretreatment.
7. Sycamore would appear to have the potential to be the most difficult of all fruits because it not only deteriorates quite quickly, but also needs a long time to pretreat and germinate. However, in practice, since it must be stored at a high moisture content and a low temperature (to maximise longevity), and at the same time, this combination overcomes dormancy, then the optimum storage conditions and optimum pretreatment conditions automatically coincide.

Details and Advice on 7 Storage/dormancy Combinations

Orthodox/hard Seeded Species

Broom (Cytisus scoparius); Scotch laburnum (Laburnum alpinum); common laburnum (Laburnum anagyroides); tree lupin (Lupinus arboreus); false acacia (Robinia pseudoacacia); and gorse (Ulex europeaus). All hard-seeded species have 'orthodox' storage characteristics, and all will require seed pretreatment to stimulate germination.

The freshly extracted, undried seeds are usually at a low enough moisture content to be treated in either of the following ways:

- o Store in a loosely-tied polythene bag in the main compartment of a refrigerator (approximately +4°C) for no more than one winter; or
- o Dry to 6-8% moisture content (fresh weight basis) which will allow sub-zero storage with little deterioration for up to 20 years. Then: pretreat artificially according to Appendices 1 and 2 and sow the following spring.

Orthodox/shallowly dormant species

Most conifers, alder (Alnus spp.) and birch (Betula spp).

Orthodox seeds all have the potential to be dried and stored, if required. It is important to note that some untreated seeds will germinate, but pretreatment will usually stimulate quicker germination of more seeds over a wider range of conditions.

Occasionally, pretreatment is harmful. For example, the germination capacity of up to 20% of commercial seed lots of Corsican pine, Scots pine and Norway spruce may be significantly reduced by pretreatment. The reasons for this are unknown but the seeds of other species appear significantly less likely to suffer damage.

The freshly extracted, undried seeds are usually at a low enough moisture content to be treated in either of the following ways:

- o Store in a loosely-tied polythene bag in the main compartment of a refrigerator (approximately +4°C) for no more than one winter; or
- o Dry conifers to 6-8% moisture content (fresh weight basis) and alders and birches to 8-10% which will allow sub-zero storage with little deterioration for up to 10 years.

Then:

- o Sow outdoors in January/February (to pretreat naturally); or
- o Pretreat artificially and sow the following spring; or
- o Sow the following spring without pretreatment.

Intermediate/shallowly Dormant Species

Fir (Abies spp.); incense cedar (Calocedrus decurrens); cedar (Cedrus spp.); Lawson's cypress (Chamaecyparis lawsoniana); and western red cedar (Thuja plicata).

On current evidence these seeds are less tolerant of drying and shorter-lived than the majority of seeds with orthodox storage characteristics. They are sensitive to desiccation, but not as much as recalcitrant seeds, or as highly perishable. The seeds exhibit the usual features of shallow dormancy in that some untreated seeds will germinate, but pretreatment will usually stimulate quicker germination of more seeds, over a wider range of conditions. Occasionally, pretreatment is harmful-especially to incense cedar.

The freshly extracted, undried seeds are usually at a low enough moisture content to be treated in either of the following ways:

- o Store in a loosely-tied polythene bag in the main compartment of a refrigerator (approximately +4°C) for no more than one winter; or
- o If storage is considered, careful drying to approximately 9-12% moisture content (fresh weight basis) will probably make sub-zero storage possible. Storage at approximately 9-12% moisture content (fresh weight basis) and -5°C to -20°C is more likely to achieve 2-5 years of storage than storage at +4°C.

Then:

- o Sow outdoors in January/February (to pretreat naturally); or
- o Pretreat artificially (according to Appendices 1 and 3) and sow the following spring; or
- o Sow the following spring without pretreatment.

Recalcitrant/shallowly Dormant Species

Horse chestnut (Aesculus hippocastanum); monkey puzzle (Araucaria araucana); chestnut (Castanea sativa); poplar

(Populus spp.); oak (Quercus spp.); and willow (Salix spp.). The combination of recalcitrance and shallow dormancy means that at the high moisture contents necessary to keep these seeds alive, any exposure to cool temperatures (-3°C to +5°C) will contribute towards dormancy breakage.

But it is vital to appreciate that the seeds of many poplar and willow species are so 'recalcitrant' that even at these low temperatures they die within days. Oaks, sweet chestnut and horse chestnut will typically decline from 90% to 50% germination over the 10-24 weeks between collection in October/November to spring sowing in March/April.

A combination of low temperature (-3°C to +5°C) to slow the rate of deterioration and minimise fungal growth plus high humidity to retard drying and adequate gaseous exchange to permit respiration are the main principles to follow (see Forestry Commission Practice Note Handling and storing acorns & chestnuts and sycamore fruits).

The most likely explanation for the earlier death of poplar and willow seeds in comparison to oaks and chestnuts is the smaller size of the former leading to faster desiccation (drying out). The freshly collected fruits are usually at a high moisture content and should be sown as soon as possible. If short-term storage is unavoidable, the following guidelines apply to a few days storage for poplar and willow seeds and a maximum of 18-24 weeks for the fruits of larger, recalcitrant species:

- Low temperatures (-3°C to 5°C) will slow fungal growth and retard respiration.
- High humidity will retard drying.
- Avoid sealed containers as sufficient gaseous exchange to permit respiration is essential.
- It may be beneficial to spray or soak fruits intermittently during storage as this appears to have a rejuvenating effect.
- Do not aim to store poplar or willow seeds any longer than a few days.

- o Do not aim to store acorns or chestnut fruits until any later than the following spring.

Then: sow outdoors as soon as possible.

Orthodox/deeply Dormant Species

Most broadleaves; juniper (Juniperus communis); Macedonian pine (Pinus peuce); and yew (Taxus baccata).

These orthodox seeds will all have the potential to be dried and stored, if required. It is also important to note that they all have an absolute requirement for pretreatment in order to bring about any germination.

However, it should be assumed that freshly collected seeds will not contain sufficient moisture to respond to pretreatment-therefore they must be deliberately imbibed (soaked and fully hydrated) before pretreatment.

The freshly extracted, undried seeds are usually at a low enough moisture content to be treated in any of the following four ways:

- o Sow immediately outdoors to pretreat naturally; or
- o Store in a loosely-tied polythene bag in the main compartment of a refrigerator (approximately +4°C) for no more than one winter, or until imbibed (soaked and fully hydrated) for natural or artificial pretreatment; or
- o Mix with moist (but not waterlogged) peat and sand and store in a loosely-tied polythene bag in the main compartment of a refrigerator (approximately +4°C) for no more than one winter; or until fully imbibed for natural or artificial pretreatment; or
- o Dry to 10-15% moisture content (fresh weight basis) allowing sub-zero storage with little deterioration for up to five years.

Then:

- o Sow outdoors, to pretreat naturally; or
- o Pretreat artificially and sow the following spring.

Intermediate/deeply Dormant Species

Norway maple (Acer platanoides); beech (Fagus sylvatica); and bay laurel (Laurus nobilis).

Seeds with intermediate storage characteristics are less tolerant of drying and are shorter-lived than the majority of seeds with orthodox storage characteristics, but not as sensitive to desiccation or as highly perishable as recalcitrant seeds. However, Norway maple (Acer platanoides) and beech (Fagus sylvatica) have two further characteristics. If Norway maple fruits are to be dried, they need very slow drying to start with and can only tolerate faster drying later.

There is also some evidence that beech nuts may have an optimum storage moisture content of approximately 15%. It appears that they deteriorate more quickly as the moisture content is progressively reduced below 15%. Otherwise the seeds in this group exhibit the usual features of deep dormancy, i.e. they all have an absolute requirement for pretreatment in order to bring about any germination. However, it should be assumed that freshly collected seeds will not contain sufficient moisture to respond to pretreatment-therefore they must be deliberately imbibed before pretreatment.

The freshly extracted, undried seeds are usually at a low enough moisture content to be treated in any of the following four ways:

- o Sow immediately outdoors to pretreat naturally; or
- o Store in a loosely-tied polythene bag in the main compartment of a refrigerator (approximately +4°C) for no more than one winter; or until imbibed for natural or artificial pretreatment; or
- o Mix with moist peat and sand and store in a loosely-tied polythene bag in the main compartment of a refrigerator (approximately +4°C) for no more than one winter, or until fully imbibed for natural or artificial pretreatment; or
- o If storage is considered, careful drying to approximately 10-15% moisture content (fresh weight basis) will

probably make sub-zero storage possible. Storage at approximately 10-15% moisture content (fresh weight basis) and -5°C to -20°C is more likely to achieve 2-5 years of storage than storage at +4°C.

Then:

- o Sow outdoors, to pretreat naturally; or
- o Pretreat artificially and sow the following spring.

Recalcitrant/deeply Dormant Species

Sycamore (Acer pseudoplatanus).

Sycamore (Acer pseudoplatanus) is unique in its combination of recalcitrance and deep dormancy. Do not let the fruits dry out or they will die. But at high moisture contents any exposure to the low temperatures (-3°C to +5°C) appropriate to storage will act as a pre-chill temperature and automatically contribute towards dormancy breakage.

The freshly collected fruits are usually at a high moisture content and should be sown as soon as possible. If short-term storage is unavoidable, the following guidelines apply: o Low temperatures (-3°C to +5°C) will slow fungal growth, retard fruit respiration and (at the high moisture content of the freshly collected seed) bring about pretreatment.

- o High humidity will retard drying.
- o Avoid sealed containers as sufficient gaseous exchange to permit respiration is essential.
- o Do not aim to store until any later than the following spring.

Then: sow outdoors as soon as possible.

OTHER THINGS TO TRY WITH DEEPLY DORMANT SEEDS

Since it may take up to two years to overcome the dormancy of some deeply-dormant fruits, you might like to try the following as a possible way of decreasing the time required.

- o Propagating the seeds from some berries without drying There is anecdotal evidence that some seeds which exhibit deep dormancy and are extracted from fleshy fruits may not require such a long dormancy breakage pretreatment-so long as seed drying is avoided. It will still be necessary to remove the seed from the fruit, thoroughly wash it (and possibly use a suitable tool to make a small hole in the seed coat to assist water uptake and embryo emergence)-but omitting a drying phase, may reduce pretreatment time significantly.

o Propagating trees with deeply dormant fruits from 'excised embryos'

In the 1930s, Florence Flemion developed techniques for carefully removing the embryos from many of the deeply dormant fruits. She found that if she placed the carefully removed embryos under light, warm, moist conditions, then, although some of the embryos decayed and died, and others merely remained firm, fresh and quiescent, within a week or two, a few embryos began to grow, green and even develop into seedlings.

The technique has since been used as a method for determining whether deeply dormant fruits are 'dead' or 'alive'- a so-called 'viability' test. But despite 50 years of development as an 'excised embryo' seed test, it has never been properly investigated as a means of propagation. So it is not known whether embryos which start to grow will continue to do so.

Perhaps they just slow down and die. Perhaps they become dwarf plants. Perhaps they need to receive the dormancy breakage pretreatment that they missed as seeds whilst they are in miniature plant form. So, try some of your own research. Have a go with species like cherry (use a vice to crack open the stone), or ash (carefully use a sharp knife to surgically remove the embryo).

Incubate the embryos on moist kitchen towel, spread them on a seed tray and enclose in a polythene bag (to retain a humid atmosphere). After a week or two, transplant any growing embryos to a pot of peat and sand and see if you can by-pass months or years of stratification (incubation of moist seeds at c. 4°C between layers of sand or soil).

VEGETATIVE PROPAGATION

Propagation is the creation of a new plant from a cutting of a branch taken from the parent plant. Many woody species found in Britain can be propagated from cuttings. Poplars and willows are the best examples and the easiest to propagate, but it is also possible to propagate apple, cherry, dogwood and hazel. However, it is important to remember that vegetatively propagated material can only have the genes of the parent plant, whereas seed material will certainly contain female genes from the mother plant and probably male genes from the pollen of a separate plant. If a wider gene base is important, seed may be the preferred route, or care should be taken to collect cuttings from a varied population or populations.

The three essential requirements for the germination of any mature, live, non-dormant seed are adequate water, good aeration and a favourable temperature.

Although light is often thought of as a further essential for germination, it is now believed to stimulate germination speed or final germination percentage by acting as a dormancy breakage agent, not as an essential requirement for the germination process itself. Light, of course is essential for the subsequent development of normal, healthy seedlings with green cotyledons (the first leaf to grow from a germinating seed) and leaves. The three main steps or stages in sowing seeds and raising seedlings are:

1. Take pretreated, non-dormant (occasionally 'chitted' or sprouted) seeds and undertake one of the following:
   - o Sow onto or into any moisture-retaining medium of loose particles such as peat, peat alternative, sand, soil, moss, perlite, vermiculite (or any desired mix of these) contained in a plant pot. Gently firm the surface. Cover with a layer of light coloured grit no deeper than 2-3 mm; or
   - o Broadcast sow onto suitably cultivated soil. Gently firm into the surface. Cover with a layer of light coloured grit no deeper than 2-3 mm; or

- o Sow in drills prepared to a depth of no more than three times the diameter of the seed. Back fill the drill.

2. Provide a suitable environment for germination either naturally (outdoors) or artificially (e.g. on a shaded window ledge, in a cold-frame, glass house or poly-tunnel). Water, ventilate and shade as necessary.

   It is important to remember how significantly temperature influences germination. For example, the seeds of deeply dormant species tend to germinate at higher percentages at lower temperatures of 10-15°C. It is also important to remember that temperatures of 20°C and above inhibit the germination of most deeply dormant seeds, and they can also re-impose a secondary seed dormancy which is just as time consuming to overcome with pretreatment as the primary dormancy. Another thing to bear in mind is that the germination of many deeply-and shallowly-dormant tree seeds also benefits from daily temperature fluctuations. Finally, ensure that the seeds are protected from mice, slugs, snails, and excessive cold, heat and damp.

3. Transplant seedlings to a wider spacing when necessary. This will result in sturdy young plants that are more likely to successfully establish in open ground when they are ultimately transplanted to their final growing position.

Ungerminated 'Failures'

Even when every possible care has been taken to handle tree seeds carefully, store them properly, pretreat them correctly and sow and nurture them optimally, some seeds apparently fail to grow. Don't give up hope-they might still only be dormant. Leave any ungerminated seeds in the ground or put pots back in the corner of the garden, and keep an eye on them for another two or three years. Many of the seeds of deeply dormant species might yet respond to natural conditions and ultimately yield seedlings.

(12)

Himachal Pradesh State Road Project

INTRODUCTION

The Himachal Pradesh (HP) has a volume of about 25,000 km of existing motorable roads including national Highways, state Highways, districts and rural roads. The HP Public Works Division (PWD) has been adopting two approaches for road construction. In the first stage, gravel road without side drains and with few structures and in the second stage, more structures, side drains and black-topping are adopted. These are basically for protecting from losses of pavement due to debris off the slopes and mode of construction involves excavation of slopes so as to level it.

The development of the hill regions in the Himachal Pradesh (HP) is impaired by difficult topography for road access. The rugged mountain topography, the complex and fragile nature of geological structures together with soft soil cover, intensive precipitation during the monsoon season and frequent earthquakes make the region vulnerable to landslides, debris flows, soil erosion and other mass wasting phenomena. Some of the important factors constraining road development in HP are large volume of soil erosion, landslides and slope failures. So, appropriate measures are necessary to preserve and protect the roads. The HP-PWD has 15 horticulture nurseries, which supply plants for shade and beautification. However, the nursery staffs are untrained local farmers. The nurseries face financial constraints because there is no separate budget for it. The seeds and cuttings are collected locally at free of costs. The program is highly successful.

The roads have puzzling features with inadequate slope protection or stabilisation. The road slopes are simply cut and are not sustainable. Moreover, the construction waste materials are tipped down the valley sides without any appropriate management (Plate 1.1). The practice of cutting vegetation along cut slopes has attributed to further scouring of the slope surface. It has pointed out a strong action for the immediate need of surface protection works. The HP-PWD is worried because there is no further vegetation works along Kalka-Simla Highway Therefore, the rationale for traffic safety engineering is highly appraised in India in recent years. As a result, demarcation of the valley side edge of the road is highly stressed.

With a view to encouraging local people in the improvement and maintenance of the Highways, the HP-PWP has recruited some 45,000 labourers. The local demand for employment led to this increment at this level. But some of these labours seem quite old. It has demonstrated that the government should properly take account the social and political interest of the local people in upgrading the road infrastructures.

Another factor that has been hindering from protecting road sides is unregulated construction of private houses close to roads. This has resulted in the blockage of culverts.

In these contexts, the HP government has a plan to introduce a state road up-gradation project for the overall economic development of the State. One of the major focuses of the project is to build sustainable environment friendly roads with appropriate application of bioengineering. The program intends to adopt a pilot bioengineering approach and replicate it based upon test and fitness of the technology.

OBJECTIVE OF THE RASK AND SCOPE OF SERVICES

Objective

The major objectives of the consultancy work are to conduct a survey on the state-of-the-art of road bioengineering along the state Highways of HP and status of institutions involved in bioengineering works in the district.

Scope of Services

The specific Scope of Services provided by the client as of November, 2006 include as follows:

1. The Bioengineering Specialist of the Consultant should familiarise him/herself with the approach to bioengineering given in the Nepal Department of Roads' "Roadside Bioengineering" manuals (1999), and outputs of the World Bank's short term consultant in bio-engineering (March 2006 and following). These are to be followed to ensure that the approach used is the same between the two strands of support to the HP-PWD.
2. Examine in detail the slopes along the first 250 km of roads scheduled for detailed design. Liaise with the Consultant's Geo-technical Specialist to put together a series of designs that combine geo-technical and bio-engineering measures to the optimum extent.
3. Undertake the same work for the second 163 km roads scheduled for detailed design.
4. Identify Sites for Pilot Trials on /along roads under the State Roads Project.
5. Detailed designs should be accompanied by specifications, digital photographs and videography and drawings of the proposed works, for all sites along the roads in question. The completed designs should be submitted to the Client for checking by the World Bank's short-term Consultant in bioengineering.
6. Demonstrative sites should be identified on some of the corridors and implementation of bioengineering works with the help of Client's staff and labour should be carried out.
7. From all detailed designs, draw together the appropriate bio-engineering items for the bills of quantities.

8. Review the experience of bio-engineering research and trials at the Dr. Y. S. Parmar University of Horticulture and Forestry, and report to the Client on its effectiveness and possible applicability to the road sector.
9. Identify training institutes and collect information on their capacities and strengths with respect to supporting the use of bioengineering.
10. Draw up lists of species appropriate for each area where bioengineering is to be undertaken on the roads, and propose practical ways in which the planting materials can be obtained or produced and appropriate to each technique that should be adopted. Where possible these should use existing HPPWP nurseries; the need for additional nurseries should be specified. The sources of materials for plant production should be specified in detail and closely related to the wide variations in altitude found in the HP road network.
11. Make recommendations on the needs of relevant staff in terms of training and other capacity support.
12. Draft an implementation strategy for bioengineering works under the two packages of detailed design roads, including works calendars and programmes; in particular, define the timings of the obligations of the PWD Horticulture Wing in producing materials for site use by the contractors. In line with this TOR, the HPPWD estimates that about 1600 kilometres of mountain road needs to be constructed within next few years to provide impetus for agriculture led growth in the next few years (Project appraisal report, PWD). The proposed roads for this are Theog-Kothkai-Hatkoti-Rohroo-Chirgaon Sandhasu, Una-Jahu-Nherchowk. Una-Aghar Barsar Jahu Bhambla Mandi, Mehatpur Una Mubarikpur Daulatpur HP boundary, Ghumarwin-Sarkaghat Jogindernagar and Kumarhatti Sarahan-Nahan (Dosarka).

This present report is based on the preliminary survey of the state-of-the-art of bio-engineeiring works at Theog-Kothkai-Hatkoti-Rohroo and Una-Jahu-Nherchowk.

THE APPROACH

Prior to the travel to Theog-Kothkai-Hatkoti-Rohroo and Una-Jahu-Nherchowk areas, documents related to the bioengineering were collected and studied. This gave the consultant to sharpen understanding on the replicable bioengineering technology relevant to the geo-morphological status of the Hindu-Kush Himalaya Region.

Then a survey checklist/guideline was prepared. Required information was collected following this. The consultant travelled several times on/along the road sides and observed the landslides. The causes of landslides with reference to the aspects of bioengineering were identified.

An in-depth survey of the available plant societies was conducted and properties of the species were studied. Species were collected to identify their potentials for replication. Based on this appropriate species were suggested.

A slope failure was identified based on the gully created by cloud outburst of 1978 and collection of red soil by the local people.

The sites for pilot trails were selected with the help of client's staff and based on the previous experience of the consultant. The detailed designs were prepared based on the observations of slope failures. The demonstration plots identified were areas having slope instability problems.

Sources of materials for plant protection were identified in terms of quantity survey along the road sides.

Discussion meetings were organised with the deans and the professors of the university and engineers of the HP-PWD and the information regarding the experiences of the organisations were collected.

Limitations of the Tasks

The limitations of the consultancy works were as follows:

This report is based on the preliminary survey of the state-of-the-art of the bioengineering works along the above stated

Highways of HP-PWD and status of bioengineering works at Dr. Y. S. Parmar University of Forestry and Horticulture and HP-PWD.

The consultant was invited for the job towards the finalisation of technical design report, which could be done after the joint accomplishment of the survey by geotechnical specialist and bioengineering specialist. Therefore, sharing of ideas in the field with the geotechnical experts regarding mitigation measures was not done.

However, the geotechnical specialist was available in Delhi Office for discussion with the bioengineering specialist

In the beginning the scope of the works was limited to identifying the sites for demonstrations while the same was extended to the establishment of bioengineering trial plots. Ultimately, it consumed more time in the construction of demonstration pilot plots through PWD labours.

Unavailability of an assistant with experiences in the identification of indigenous plants of local names took more time in the survey. Similarly, lack of field demonstrations with respect to dealing with labour was supplemented by an experienced person from Nepal.

Lapse of time due to arrangement delayed the works.

Plate No. 1.1: Excess and improper dumping of excavated material

CONCEPT, APPROACH AND PRACTICABLE TECHNIQUES OF BIOENGINEERING

Wood and living plants were the only materials for hill and slope stabilization works for hundreds of years. These days some of the old techniques have been modified and applied, which mainly use live material such as willow branches, willow cuttings, and rooted deciduous trees. The duration for implementing vegetation for slope stabilization works depend on the technique of soil bioengineering measure and the type of the plant that has been used. The techniques used in soil bioengineering are traditional and have been used since very long time ago. The first use is recorded from the 5th century in ancient China. Soil-bioengineering methods were used by the Romans and in the Middle Ages.

Shiechtl documented these techniques and from 1950 onwards a new start of soil bioengineering began. John H. Howell's works has been pioneer in roadside bioengineering of Himalayan regions.

Bioengineering is a sustainable approach where the combination of live and dead plants and plant parts are used as building materials for erosion reduction and upland slope prevention. This is based on engineering concept. The combination of vegetative engineering systems and civil engineer__g systems normally offers the most complete natural and cost effective solution to the variety of instability problems affecting a site

The overall objective is to provide better living conditions to the population by the provision of environment sound and improved sustainable mountain road highway system so as to make a rapid economic growth in tune with economic growth of the country. Specifically bioengineering is adapted with a view to

i) examine soil bioengineering needs of project,

ii) monitor existing road system regarding their slope stabilization works.

iii) study plant species compositions and plant species which can be used for future soil bioengineering purpose.

iv) survey a plan of soil bioengineering activities.

v) test plant species in combination with different soil-bioengineering techniques for its suitability in winter plantations.

vi) stabilize the unstable slopes through a suitable application of vegetation in different soil-bioengineering methods, either alone or in combination with civil engineering structures.

vii) use indigenous materials, low capital costs compared to civil engineering structures.

Road bioengineering is totally labor based approach providing benefits to the local people. It focuses on minimizing the damage to the natural environment as well as reducing the future maintenance costs. Major portion of the cost of the road construction goes toward the payment of wages to the local labor. Road construction approach incorporates social, environmental and technical issues. Due to these facts, it is getting increasingly prioritized by national and international organizations in hill road construction. The people's participation is most essential for the success of slope conservation works. The use of local materials motivates local people's participation. Soil bioengineering is an important and an integral part of the road construction, which is environment sound and cost effective. The use of plants for land stabilization is applied either alone, or in conjunction with civil engineering structures and non-living plant materials to reduce shallow-seated instability and erosion on slopes.

Bioengineering Approach Adopted by the Nepal Department of Roads

Experiences have proven that without adequate planning roads are not sustainable. Evidences have also shown that in the past some hill roads of the Himalayan region needed rehabilitation because of no maintenance or very poor maintenance making the road a very expensive project. The conventional method of road construction are generally being followed and are increasingly being questioned due to the use of large scale engineering structures with high investments.

The manual of Road Department of Nepal Government has been developed from an enormous amount of experience throughout the road network of Nepal between 1984 and 1998. The manual emphasizes on an Environment Friendly Road (EFR) construction, the use of locally available resources, manpower, material, technologies, skills and local political situation. The manual is based on the tested techniques and exceptional geomorphology, steep slopes, intensive rainfall and restricted economy of the country. This current survey was conducted according to the guidelines of the manual in conformity with the Scope of Services given in chapter.

Practicable Ways for Technique Adoption and Construction of Nurseries

Bioengineering methods can be used to protect almost all type of slopes against erosion and reduces the instance of shallow failures. However, the establishment process may take some time. Vegetation is more dynamic than inert and takes some time to reach their maximum strength. They tend to become stronger over time. The application of soil bioengineering in combination with civil engineering structures would be most appropriate. Dry stone walls or gabion retaining structures are built in the valley side of the road to protect the fill volume to reduce mass wasting and erosion process. Erosion control plantings are carried out on bare fill surface. Every attempt is made to maintain mass balance of road cut and fill. The incorporation of soil bioengineering is a vital and integral part of road construction.

Cuttings and rooted plants are used during the dormant season. Various methods are available for hill side slope stabilization.

The methods for seeding are dry-seeding and hydroseeding. The seed will be protected with straw in combination with bitumin or meshes of jute and wire on exposed areas. The stabilisation can be increased through transplanting stump sprout deciduous trees after revegetation with seeding.

Loose rock slopes can be stabilized with different types of brush layering. The application with rooted plants are known as hedge layer. The unrooted plant installation are brush layer. A combination of both is a hedge brush layer.

Dewatering and stabilization of wet slopes can be done through the use of drain and slope fascines. The fascines consist of live branches of willows that are tied together with wire.

There are numerous different hill side slope stabilization methods, which utilize plants in combination with construction of wood, stone and wire such as planted pole walls, live slope grids, live bamboo crib walls, vegetated stone walls and vegetated gabions. Choosing the right method depends on various factors such as the position of slope, ground and available material. The sections given below presents the techniques adopted in brush layer, drainage fascines, pallisade, single tree planting and grass planting. The advantages and disadvantages of each of these techniques are also given under the respective techniques.

Brush Layer Techniques

Brush Layer consists of placing live branch cuttings laid in small benches excavated into slopes. The benches can range from 2 to 3 ft wide. Live materials should be 1/2 to 2 inches in diameter. It is long enough to reach the back of the bench. The side branches should remain intact for installation.

Plate No. 2.1 : Brush layer construction placing live branch cuttings in small benches into slopes

The cuttings are placed into slopes and oriented perpendicular to slopes. This is a more effective for earth reinforcement and mass stability of slope. It is also recommended on slopes up to 2:1 in steepness. It serves as tensile inclusions or reinforcing units. It assists in retarding runoff and reducing surface erosion.

The construction breaks up the slope length into series of shorter slopes separated by rows of brush layer. It also reinforces the soil with un rooted branch stems and reinforces the soil as roots develop, adding significant resistance to sliding or shear resistance.

Whereas on dryer sites it helps infiltration on dry sites and drying excessively wet sites. It also adjusts the sites micro-climate, thus aiding seed germination and natural regeneration. The technique also acts as horizontal drains by redirecting and mitigating adverse slope seepage.

The construction is started at the toe of slope. Benches are excavated horizontally, on the contour, or slightly down the slope, if needed to aid the drainage. The surface of the bench slope should be constructed in such a manner that the outside edge is higher than the inside part. The live branch cuttings placed on the bench. Branch growing tips (shoot bud) should be aligned towards the outside of the bench. The backfill is placed on top of the branches and compacted to eliminate air spaces. The brush tips should extend slightly beyond the fill to filter sediment. Each lower bench is backfilled with soil obtained from excavating the bench above.

Drainage Fascines

Fascines are used for a variety of slope stabilization purposes. Fascines slow runoff, catch debris and reinforce the slope due to rooting. Drainage fascines in particular stabilize and drain slopes and are built into rills or small gullies.

Immediately after establishment they have a draining effect because the water is channelled through the straight branches. With shoot and root development they form a strong line of vegetation. Additionally they achieve water-removal due to transpiration of plants. (HOWELL, 1999a; SCHIECHTL, 1992; SDC, 1990)

The bundles are tied together with 1,5 -2,0 mm wire every 1 m and fixed into the grnund every meter by wooden pegs or live stacks with a minimum length of about 80-100 cm. For the lower third of the fascine green branches of non-sprouting plants or gravel can be used. After placement the fascines are covered with a thin soil layer of about 3-4 cm. All branches should be in contact with soil to enable rooting and shoot development. (FLORINETH and RAUCH, 2001; SCHIECHTL, 1992)

Advantages and Disadvantages of Drainage Fascines

It is a fast and simple construction method, which is effective even after silting. Drainage fascines provide drainage immediately after establishment (SCHIECHTL, 1992). The disadvantages of this technique are as follows. A large amount of straight and long plant material is needed. They can only drain a limited amount of water, up to 5 litres per second. Otherwise tile drains have to be used. (FLORINETH and RAUCH, 2001; SCHIECHTL, 1992).

Palisades

Vegetated palisade constructions are ancient techniques, which were already used by the first people in Central Europe. At the beginning of the 20th century vegetated palisade constructions were still common in Bohemia, Moravia and Slovakia.

Palisades are used for protection of small but deeper, narrow gullies and shallow V-shaped rills. Immediately after construction they provide mechanical protection by catching debris, armouring and reinforcing gully floors. These effects are increasing after shoot development of cuttings. As a result of plant transpiration, they achieve water-removal from soil. (FLORINETH and RAUCH, 2001; SCHIECHTL, 1992).

Huge live poles are driven or dug half of their length, vertically into the ground. 5 to 20 live poles per running meter with at least 5 cm in diameter should be used.

On their top they are fixed by wire on one or two cross beams, which have been anchored into the sides of the gully. The distance between the palisades depends on the steepness of the gully slope and profile of the gully floor. On slopes with less than 30° inclination 2 m are convenient, on slopes with an inclination between 30° and 60° a distance of 1 m is recommended. (FLORINETH and RAUCH, 2001; HOWELL, 1999a; SDC, 1990)

Advantages and Disadvantages of Palisades

Palisades are an effective low-cost structure, require a small amount of plant material, are fast to establish and cause little site disturbance during construction. The disadvantages of the palisades are that they are only suitable for narrow and not too deep gullies with a maximum width of about 5 m and a maximum height of 2 to 4 m and only sporadic water regime. (SCHIECHTL, 1992; SDC, 1990).

Single Tree Planting

Growing shrubs and trees form a dense network of roots in the soil. Reinforcing and in the longer term anchoring, are the main engineering functions. Shrub and tree planting can be applied on almost every slope. It is often used in combination with other soil-bioengineering measures. (HOWELL, 1999a; SDC, 1990).

Rooted plants from the nursery (usually raised as polypot seedlings) are planted in off-set rows or in other specific pattern on the slope. One plant per m^2 should be enough. Main considerations are the costs of establishment and the period in which a dense plantation is required.

The plants are put into a pit large enough for the rooting system. After filling soil carefully around the cylinder of roots and soil from the polypot, the soil is firmed gently around the plant. Stump-sprouting species should be preferred. Mulching, with for instance compost or chopped grass helps the young plants to grow faster by regulating the moisture and suppressing competing vegetation. (HOWELL, 1999a; SCHIECHTL, 1992; SDC, 1990)

Advantages and Disadvantages of Single Tree Planting

Due to interception, single tree plantations provide a good surface protection. The disadvantages of the single tree plantings are that plants have to be carefully raised and selected in the nursery. Shrub and tree plantations take about 5 years to contribute significantly to slope strengthening and require care and protection in the first 3 years (HOWELL, 1999a; SDC, 1990).

Grass Planting

For grass planting, grass slips, cuttings, or clumps grown from seed are planted in lines or at random on the slope. Grass plantations protect the slope, due to rooting and by providing a surface cover. They reduce the speed of runoff, catch debris and armour the slope. (HOWELL, 1999a)

By using a planting bar holes just big enough for the roots are made. The grass slips or cuttings are placed into it, the soil filled around it and firmed gently. Grass planting is started from the top of the slope working downwards.

Different patterns of grass lines can be used as given below:

Contour/horizontal Lines

They reduce speed of runoff and catch debris thereby armouring the slope.

Down-slope/vertical Lines

Armour, reinforce and drain the slope. Used for damp sites and poorly drained materials where an intensified infiltration can lead to liquefaction of the soil.

Diagonal Lines

Main functions are armouring and reinforcing. Secondary they catch debris and drain the slope. Best compromise of horizontal and vertical planting. Random planting: it is often used in combination with jute netting on very steep harsh slopes where complete surface protection is needed.

EXAMINATION OF SLOPES AND IDENTIFICATION OF SITES FOR PILOT TRIALS

The roads are located over the middle Himalayan range with steep slopes and V-shaped river valleys. The side slopes are as steep as 10 degrees up to over 50 degrees. The V-shaped valleys characterize old and newer landslides areas and steep river gullies. This situation has constrained road construction requiring the cutting of hill sides for its completion.

The geomorphology of the road areas is influenced by the narrow river valleys and its tributaries, hills with mild to steep slopes and colluvial deposits at the foot of the slopes. The slopes range from 10 degrees to 65 degrees on both sides of the river valleys. The area is highly erosion prone due to the steepness in general, old land slides debris and thick colluvium's deposits on hill slopes and rock slide debris deposits along the slopes.

Moreover, human interference such as faulty agricultural practices, improper methods of road construction and deforestation has further accelerated the whole process. The road construction has also contributed to numerous small slides and soil creeps at different stretches of the road alignment. Erosion forming deep rills and cuts, effects of torrential monsoon rains and rock mass failure are given in Plates 3.1, 3.2 & 3.3 respectively. Plates 3.4, 3.5, 3.6 and 3.7 present slides requiring various bioengineering treatments.

Plate No. 3.1: Erosion in Bhaugoli nala, Theog road forming deep rills and cuts damaging road surface attributing to stream bank erosion forming different size rills. The water discharges over road surface.

Plate No. 3.2 The torrential monsoon rainfall has resulted into transportation of materials and trees down stream at Bhaugoli nala

Plate 3.3 Rock mass failure in densely vegetated slope which could affect the road consisting of Quarzitic, phylites rock materials at km 58+450

Plate 3.4: A fresh slide due to saturation after intensive rainfall in a highly weathered shale in Una-Nherchowk road (approx 12.2 km from Jahu to Nherchowk)

Plate 3.5: A slide under pine forest in steep angle of cut, combination of bioengineering with civil structure required.

Plate 3.6: A slide over highly disintegrated underlying rock-mass (10 km from Jahu-Una).

Plate 3.7: A moving slide and structure is required at toe of slope

Identification of the Sites for Pilot Trials and Demonstrative Plots

Theog road is a hilly road. It starts from Theog, which is a ridge point and ends at valley bottom in Rohru. The altitude varies from 1200 M to 2500 above mean sea level.

The vegetation zones can be classified into three categories

a) 1000 to 1700 m a.s.l.,

b) 1700 m to 2400 m a.s.l. and

c) 2400 to 2800 above m a.s.l.

The different sections of the surveyed road areas have dominant species such as Pines roxburghii, Pinus wallichiana, Cedrus deodara and Quercus spp., Robenia psuda and Alnus nepalensis.

The associate species are Salix spp., Morus spp. Salix spp. Robinia spp., Ailenthis, Albezia spp., Populus spp. Rhododendron arborium, Abies pindrow, Picea smithians, Psuga dumosa etc.

The dominant shrub species are Berberis aristata and the associate species are Rubus ellepticus, Rubus spp., Artemeris vulgaris, etc.

The grass species commonly seen are cymboggon spp. (khar), Penisetum purpurium, Baghad etc.

Identification of the Sites for Pilot Trials and Demonstration Plot

Five different sections along the road alignment were identified for bioengineering pilot work. Sundali nala was considered to be the most appropriate site for construction for November. The trial plot proposed for construction was to be established in a natural manner, without further watering. The bioengineering details are provided.

Plate no 3.8: Selected trial plot for bioengineering construction in Sundali nala The methods adopted in the construction demonstrative plots are as given below:

Four different types of techniques; namely, brush layering (Plate no. 3.9 & 3.10), fascines (Plate no. 3.11), vegetated pallisades construction (live check dam) (Plate No. 3.12) and bamboo crib wall (Plate no. 3.13) are in use in the demonstration plots: The cost estimation is provided in Table No. 4 in the annex.

Cuttings and rooted plants are used during the dormant season. The methods for seeding are dry-seeding and hydroseeding. The stabilisation can be increased through combination of small scale civil engineering works with appropriate bioengineering techniques.

There are numerous different hill side slope stabilization methods, which utilize plants in combination with construction of wood, bamboos, stone and wire such as planted pole walls, live slope grids, live wooden crib walls, live bamboo crib wall vegetated stone walls and vegetated gabions.

Choosing the right method depends on various factors such as the position of slope, ground and available material.

Plate No.3.9: Brush layer construction trial plot in Sundali nala

Loose rock slopes can be stabilized with different types of brush layering. The application with rooted plants are known as hedge layer. The unrooted plant installation are brush layer. A combination of both is a hedge brush layer. Dewatering and stabilization of wet slopes can be done through the use of drain and slope fascines. The fascines consist of live branches of willows that are tied together with wire. The picture shows the vegetated pallisade construction (live check dam) at Sundali nala

Plate No. 3.10: Willow cuttings laid for Brush layer construction

Plate No. 3.11 Fascine consisting of live branches of willows tied together with wire.

Plate No. 3.12 vegetated pallisade construction (live check dam) at Sundali nala.

Plate No. 3.13 Bamboo crib wall construction in Sundali nala

Useful Plant Species for Bioengineering Works

During the survey visits, various plant species along road alignments were observed. Particularly, tree species such as Acacia catechu, Albizzia lebbek, Albizzia stipulata, Bauhinia variegata, Bombax ceiba, Ficus religiosa, Ficus roxburgii, Grewia oppositifolia, Mallotus phillipensis, Mangifera indica, Melia azadiraeh, Morus alba, Salix sp. Phanera rahlil, Phoenix sylvestris, Phyllantus emblica, Pinus roxburghi, Pistaua integrimma, Prunus cerasoides, Rrobinia pseudoacacia, Sapium spp. Terminalia belerica, Toona cilia/cedrela toona, Utis spp and Zizyphus jujube were noted along the road. These tree species provide raw materials for wood industries, medicines, timber, fodder, shed, firewood and agriculture tools.

Likewise, the hills have a number of shrub species such as Achyranthes bidentata, Adhethota vasica, Agava Americana, Aromatic plants, Azadirehta indica, Barberis asistate, Calotropis procora, Cassia opaca, Flocourtia indica, Jatropha spp. Lantana camera, Murraya koenigii, etc.

These shrub species are the sources of fencing, firewood, foliage, green manure, medicines, flowers, insecticides, food, etc. Some grass species include Ageratum conyzoides, Arundinaria spp., Cymbopogon, Cynodon dactylon. The grass species are also useful for thatch, fodder, pasture and medicines.

STATUS OF BIOENGINEERING WORKS AT DR. Y.S. PARMAR UNIVERISTY OF HORTICULTURE AND FORESTRY AND HP-PWP

The Dr. Y.S. Parmar University of Horticulture and Forestry is a pioneer institute for bioengineering research, particularly in the fields of agro-forestry and horticulture. The university is equipped with professionals of high academic standards and technical expertise.

It possesses potentials for bioengineering works as it implements reforestation, which has direct relationship with bioengineering. Lots of efforts were put for the stabilization of Highways through its efforts. A large variety of ornamental plants were planted for beautification along the Highways. The beautification works are highly appraised. The works are mainly plantation of seedlings and stump plantings. The plant species can however, be used for bioengineering works along the Highways.

Technical Capacity of the Dr. Y.S. Parmar University of Horticulture and Forestry

Dr. Y.S. Parmar University of Horticulture and Forestry is well equipped with all sorts of forestry experts and infrastructure. There is a huge collection of forestry related literatures in the university library. The university produces Horticulturalist. Its product has a good market demand. Almost all the horticulturalists in PWD are the product of the Dr. Y. S. Parmar University of Horticulture and Forestry.

Constraints

- o There is no curriculum in the university syllabus, which focuses on bioengineering techniques for hill slope stabilization.
- o There is no expert specialized in bioengineering technique in the university.
- o The university does not have adequate physical facilities and funds to launch bioengineering related academic program and trainings.
- o There is a lack of bioengineering related literature in the university library.
- o As bioengineering involves different stakeholders, there is a need of integrating sociological and economic aspects in the university.

Technical Manpower of Himanchal Pradesh Public Works Division (HP-PWD)

Himanchal Pradesh is categorized into three zones for road construction purpose.

The PWD is the main entity to look after the road network development tasks in these zones. The division needs technical manpower as per its road construction requirement in each of these zones. A cursory look at the technical manpower employed under the Himanchal Pradesh Public Works Division (HP-PWD) reveals that it has a plenty of technical expertise.

As shown in the table below, it employs a total of 1,357 engineers. The PWD engineering section is headed by an Engineer in Chief. Four Chief Engineers are working under the Engineer in Chief. Each Chief Engineer has 4 Superintendent Engineers and in the same manner a Superintendent Engineer has 4 Executive Engineers. Furthermore, each Executive Engineer has 4 Assistant Engineers working under him. Similarly, all Assistant Engineers have 4 Junior Engineer as subordinates.

Table : Number of technical staff of PWD Category of technical staff

Number	Percent	Head engineer of PWD
1	0.1	Chief engineers
4	0.3	Superintendent engineer
16	1.1	Executive engineer
64	4.7	Assistant engineer
256	18.9	Junior engineer
1016	74.9	Total engineers
1357	100	Labor

Source: Discussion with PWD.

PWD has a horticulture wing with 9 engineers of which, 4 are Assistant Engineers and 5 Junior Engineers. However, there are eleven more vacant positions which need to be filled in for the post of junior Engineer Horticulture. Of the above figures, Assistant Enginners Horticulture consists of 0.3% of the total professional manpower and Jr. Engineers contain 0.36% of the total professional manpower in the PWD. The PWD also employs 45,000 labors.

Horticulture Wing of PWD

The horticulture wing of the PWD was established in 1977 with the objective of road soil conservation and beautification through ornamental and non timber plant species.

Initially two junior horticulturists and a few horticulture inspectors were deputed to the PWD from Himalchal Pradesh Horticulture Department. One DHO was also deputed as an assistant engineer.

The first major assignment was to re-vegetate NH-22 highway through plantation of mainly ornamental plants. The

NH-22 consists of mainly Kalka to Shimla road (84 km) and Ambala-Parwanoo-Solan to Shimla road (62 km).

In 1981 two more junior engineers were recruited for plantation works of NH-22 highway. With the above given manpower the NH-22 highway was made green. The main works executed were single row plantation, double row plantation, group plantation and complete hedge plantation on edge. Most parts of the works were carried out on valley side.

Soil binding plant species were selected and planted on hillside slope through stump planting method. Different grass seed species were also sown. The main ornamental shrubs planted applied were: Jashmin sp., Bahun behiliya sp., spinis brun (yellow flower).

An additional horticulture sub division was created at Bilaspur in 1988 for the plantation and conservation works of NH 21 highway (Chandigardh to Manali road). Similarly, sub division horticulture was established at Rampur Bushahar for NH-22 road in 1994. One more sub-division horticulture was created at Pandoh on NH 21 for the same.

At present, there are nine professional staff consisting of four assistant horticulture engineers and five junior engineers horticulture working in HP mainly in the above mentioned four sub divisions. Currently, eleven posts of junior engineers are lying vacant in the HP state. A junior engineer is qualified BSc agri horticulturist by profession.

Constraints and Needs HP-PWP and Horticulture Wing

The constraints of HP-PWD in the field of bioengineering and its technique, revealed from the field visits and discussion with the engineers, are as follows. o There is a lack of technical manpower in Horticulture wing. There is a vacancy of 11 Junior Engineers Horticulturalist in the division. The wing needs more technical manpower with expertise in bioengineering techniques. o The Horticulture wing looks small in its scope and it needs further expansion. The tasks of the Horticulture section needs to be expanded in each of the road construction zones of the PWD.

- o No engineer has an academic qualification in horticulture.
- o The labors lack training in bioengineering works.
- o The engineers working under PWD lack exposure to bioengineering techniques.
- o There are a very scanty number of horticulturalists in the division.
- o The Junior Engineers supervise the technical works and prepare bills. They have to be really involved in bioengineering construction sites during survey of the sites, construction, repair and maintenance of the roads. However, they lack background knowledge and skills of bioengineering related techniques.
- o The PWD horticultural wing could organize bioengineering related training programs for the engineers, but they lack sufficient exposure.

CONCLUSION, RECOMMENDATIONS AND IMPLEMENTATION STRATEGY

The HP government has a plan to introduce a state road up-gradation project for the overall economic development of the State. One of the major focuses of the project is to build sustainable environment friendly roads with appropriate application of bioengineering. The program intends to adopt a pilot bioengineering approach and replicate it based upon test and fitness of the technology.

In these contexts, the major objectives of the consultancy work were to conduct a survey on the state-of-the-art of road bioengineering along the state Highways of HP and status of institutions involved in bioengineering works in the district.

Specifically, the Scope of Services for this consultancy included the review of the concept of bioengineering and its approaches, identification of applicable techniques of bioengineering, examination of slopes along 250 and 163 kms roads, preparation of detailed designs with specification, identification of pilot trials and demonstration plots, drawing of appropriate items for bills of quantities, review of the bioengineering research experience of Dr. Y.S. Parmar University of Horticulture and Forestry and HP-PWP, technical capacities of the university and the HP-PWP, recommendations on enhancing the capacities of the training institutions and implementation strategy for the bioengineering works along HP State Highways.

This present report is based on the preliminary survey of the state-of-the-art of bio-engineeiring works at Theog-Kothkai-Hatkoti-Rohroo and Una-Jahu-Nherchowk.

The survey approach included collection of documents and travel to the Theog-Kothkai-Hatkoti-Rohroo and Una-Jahu-Nherchowk areas. A survey checklist/guideline was prepared and implemented to collect the required information. The consultant travelled several times on/along the road sides and observed the landslides.

The causes of landslides with reference to the aspects of bioengineering were identified. An in-depth survey of the available plant societies was conducted and properties of the species were studied. Species were collected to identify their potentials for replication.

Based on this appropriate species were suggested. Slope failures were identified based on the gully created by cloud outburst of 1978 and collection of soil by the local people. The sites for pilot trails were selected based on the previous experience of the consultant. Detailed designs were prepared accordingly.

Sources of materials for plant protection were identified in terms of quantity survey along the road sides. Discussion meetings were organised with the deans and the professors of the university and engineers of the HP-PWD and the information regarding the experiences of these organisations were collected based on these. Several photographs and videography were taken during the survey.

The major limitation of the consultancy work was that it is based on the preliminary survey of the state-of-the-art of the bioengineering works along the above stated Highways of HP-PWD and status of bioengineering works at Dr. Y. S. Parmar University of Forestry and Horticulture and HP-PWD.

The major findings of the survey are as follows:

The road section from Theog to Rohru is located at steep slopes. The area is highly erosion prone due to the steepness, old land slides debris and thick colluvium deposits on hill slopes and rock slide debris deposits along the slopes. Faulty agricultural practices, improver methods of road construction and deforestation have accelerated the erosion.

The survey has identified five sites along the road alignment for demonstration plots. The techniques applied in the demonstration plots included bamboo crib wall, bush hedge layer, fascine, palisade and grass plantation.

Detailed designs were prepared and the specifications presents applicable rules for the collection of seeds (grass, tree, shrub), cuttings (grass, tree, shrub), nursery construction and establishment, nursery operation and management, nursery production of grass, trees and shrubs and hardwood plants, extraction of plants from nursery, slope preparation for planting, site planting and sowing, brush layering, palisades, fascines and protection of sites.

Those areas which have been identified earlier as trial plots were discussed during the presentation held on 23rd June 2007 as part of the presentation on " Road side Bio Engineering for HP state Roads Project. Even though initially there was some reservations expressed on the intended site selection, such as Bhaugolinala, the participants from HP PWD appraised this particular site has problem, in case if this site is not treated it may threaten the new bridge that was constructed recently.

Hence the consultants understand that this site shall remain as the trial plot as per the original recommendation followed by affoestation by forest department in future. Bioengineering works have been proposed in all areas along roadside and slopes that can affect the road irrespective of the different types of land use system such as forest land.

The report also presents drawings of appropriate bioengineering items for the bills of quantities. It also presents estimates cost for bioengineering trail plot. Though the Dr. Y. S. Parmar University of Horticulture and Forestry is a pioneer institute in HP, but its research and trial activities have not focussed on bioengineering techniques.

Similarly, the Horticulture wing of HP-PWP has been focussing on the production and plantation of ornamental plants rather than the plants suitable to bioengineering works. Both of these institutes could play a significant role in the research and application of bioengineering techniques in HP, but their roles seem inadequate for the purpose.

Dr. Y.S. Parmar University of Horticulture and Forestry is well equipped with all sorts of forestry experts and infrastructure. There is a huge collection of forestry related literature in the university library.

The university produces Horticulturalist. Its product has a good market demand. Almost all the horticulturalists in PWD are the product of the Dr. Y. S. Parmar University of Horticulture and Forestry. But the university has facing a number of constraints such as: There is no bioengineering related curriculum in the university syllabus.

There is no expert specialized in bioengineering technique in the university. The university does not have launch bioengineering related academic program and trainings. There is a lack of bioengineering related literature in the university library. As bioengineering involves different stakeholders, there is a need of integrating sociological and economic aspects in the university.

There are nine professional staff consisting of four assistant horticulture engineers and five junior engineers horticulture working in HP-PWP. Currently, eleven posts of junior engineers are lying vacant in the HP state. A junior engineer is qualified BSc agri horticulturist by profession.

The HP-PWP and Horticulture Wing have been facing a number of constraints such as there is a lack of technical manpower in Horticulture wing. There is a vacancy of 11 Junior Engineers Horticulturalist in the division. The wing needs more technical manpower with expertise in bioengineering techniques.

The Horticulture wing looks small in its scope and it needs further expansion. The tasks of the Horticulture section needs to be expanded in each of the road construction zones of the PWD. No engineer has an academic qualification in horticulture. The labors lack training in bioengineering works.

The engineers working under PWD lack exposure to bioengineering techniques. There are a very scanty number of horticulturalists in the division. The Junior Engineers supervise the technical works and prepare bills.

They have to be really involved in bioengineering construction sites during survey of the sites, construction, repair and maintenance of the roads. However, they lack background knowledge and skills of bioengineering related techniques.

The PWD horticultural wing could organize bioengineering related training programs for the engineers, but they lack sufficient exposure. The engineers lack exposure on integrating social factors in the survey of the roads, its design and construction.

Recommendations

Training Needs of the Relevant Staff of HP-PWP and other capacity support Based on the above-stated findings following training needs have been identified for the relevant staff of the HP-PWP.

1. Recruit more horticulturalists with sufficient background in bioengineering.
2. Recruit an Executive Engineer as chief of the horticulture wing of the PWD so as to have to good image among the engineers in the division and smooth operation of bioengineering works in the wing.
3. Establish a training wing in the Horticulture wing of PWD.

4. Develop a training curriculum for the different categories of the engineers of PDW and labor in collaboration with the Dr. Y.S. Parmar University of Horticulture and Forestry.
5. Provide on the job trainings to the different categories of engineers of PWD in bioengineering.
6. Organize intensive workshops for executive engineers and assistant engineers imparting knowledge on different techniques of bioengineering them.
7. The Horticulturists and engineers need training on social mobilization/motivating people in bioengineering works.
8. Organize intensive trainings for gang leaders of the labors to equip them with skills suitable to application of bioengineering techniques in hill road construction, repair and maintenance.

Training Needs of the University and other capacity support.

Following recommendation have made to improve the technical capacity of Dr. Y. S. Parmar University of Forestry and Horticulture.

1. There is an urgent need of starting bioengineering related curriculum in the university.
2. The university should have experts specialized in bioengineering.
3. The university should launch a separate intensive training to the manpower working under it. The university requires additional physical facilities and funding for this.
4. As bioengineering involves different stakeholders, there is a need of integrating sociological and economic aspects in the university. For this, the university needs a sociologist and an economist to provide trainings on social mobilization and integrating rural poor in road survey, design and construction.

5. Workshops for Superintendent Engineers, Executive Engineers need to be conducted.
6. Similarly, a training workshop is recommended for some Executive Engineers and mainly Assistant engineers. They play an important role in decision making.
7. A more detailed technical training should be designed for Junior Engineers, who are in field more frequently in the construction, monitoring, and supervision activities.
8. Gang leaders (mate) are the field workers also supervising the workers in the workplace. They play a vital role in working as well as demonstrating the works to workers. They will need on the job training for about 7days, where they will apply some bioengineering techniques for slope stabilization and erosion control.

Implementation Strategy for Bioengineering Works

- o Recruit a bioengineering expert for 16 moths I two years time under PWD Project to work inside the PWD so as to work in close coordination with superintendent and executive engineers and be involved in providing advisory services for training and implementation part. Consultant should be recruited initially for a period of 4 months for advisory works in preparation phase. Such consultant should be responsible to prepare detailed specific designs and implantation plans. Then, consultant can be recruited for 16 months in a phased manner; 8 months in the first year and 8 months in the second year.
- o Develop training package for Horticulturalist and organize bioengineering trainings for them.
- o Develop workshop training materials and organise workshops and trainings for superintendent engineers and executive engineers.
- o Develop reading materials for workshops of engineers. Arrange workshops for chief engineers, executive

engineers, superintendent engineers and junior engineers.

- o Develop training packages for gang leaders and provide training to them. Orient the labours through trained gang leaders.
- o Establish at least 7 nurseries for Una and Theog roads.
- o Plant seedlings at roadsides.
- o Develop separate package of bioengineering techniques for dormant and monsoon seasons.
- o A socio-economic should be recruited and he/she should be made responsible to launch social mobilisation to raise local people's awareness on protecting lands from landslides.

Bibliography

Aukland, L.: (1997). *The Effect of Selective Logging on Amphibian Diversity of Budongo Forest*, Uganda. B. A., Oxford University.

Babweteera, F.: (1998). *Influence of Gap-size and Age on the Diversity and Abundance of Climbers in Budongo Forest Reserve*, Uganda. M. Sc., Makerere University.

Bahati, J.: (1995). *Impact of Arboricidal Treatments on the Natural Regeneration of Species Composition in Budongo Forest*, Uganda. M. Sc., Makerere University.

Barrows, M.: (1996). *A Survey of the Intestinal Parasites of the Primates in Budongo Forest*, Uganda. Glasgow University.

Bennet, H.: (1996). *Study of Kinship, Social Relations and Possible Avoidance of Incest among Budongo Chimpanzees*. M. A., Edinburgh University.

Bethell, E.: (1998). *Vigilance in Foraging Chimpanzees*. University College, London University.

Fairgrieve, C.: (1995). *The Comparative Ecology of Blue Monkeys (Cercopithecus mitis stuhlmanni) in Logged and Unlogged Forest, Budongo Forest Reserve*, Uganda, Edinburgh University.

Hills, C.: (1998). *An Analysis of Phylogeny and Gene Flow in a Population of Pan troglodytes schweinfuthii*. M. Sc., Oxford University.

Johnson, K.: (1993). *Local Uses of Budongo's Forest Products*. M. Sc., Oxford University.

Johnson, K.: (1996). *Hunting in the Budongo Forest*, Uganda. SWARA.

Johnson, K.: (1996). *Local attitudes towards the Budongo Forest*, Western Uganda. Indigenous Knowledge & Development Monitor.

Miller, P. L.: (1993). *Some dragonflies of the Budongo Forest*, Western Uganda. Opusc.

Mork, K. & Solvang, R.: (1996). *Edge Effects on an Understorey Bird Community in Budongo Forest Reserve*, Uganda, Agricultural University of Norway.

Musamali, P.: (1996). *The Ecology, Diversity, and Relative Abundance of Small Mammals in Primary and Disturbed Forest Compartments in the Budongo Forest Reserve*. M. Sc., Makerere University.

Owiunji, I.: (1996). *The Long Term Effects of Forest Management on the Bird Community of Budongo Forest Reserve*, Uganda. M. Sc., Makerere University.

Reynolds, V.: (1965). *Budongo: A Forest and its Chimpanzees*, London, Methuen.

Reynolds, V. & Reynolds, F.: (1965). *Chimpanzees of the Budongo Forest*, New York, Rinehart and Winston.

Shiel, D.: (1997). *The Ecology of Long-term Change in a Ugandan Rainforest*. D. Phil., Oxford University.

Spini, L.: (1998). *The Frequency and Significance of Social Grooming in the Sonso Community of Chimpanzees at the Budongo Forest*, Uganda. M. Sc., Oxford University.

Index

E

F

G

H

I

K

L

T

U

V

W

□□□